T0340096

ARTIFICIAL INTELLIGENCE AND MACHINE LEARNING IN SMART CITY PLANNING

ARTIFICIAL INTELLIGENCE AND MACHINE LEARNING IN SMART CITY PLANNING

Edited by

VEDIK BASETTI
*Department of Electrical and Electronics Engineering,
SR University, Warangal, Telangana, India*

CHANDAN KUMAR SHIVA
*Department of Electrical and Electronics Engineering,
SR University, Warangal, Telangana, India*

MOHAN RAO UNGARALA
*Department of Applied Sciences, Université du Québec á
Chicoutimi, Chicoutimi, QC, Canada*

SHRIRAM S. RANGARAJAN
*Department of Electrical and Electronics Engineering,
Dayananda Sagar College of Engineering, Bengaluru,
Karnataka, India*

ELSEVIER

Elsevier
Radarweg 29, PO Box 211, 1000 AE Amsterdam, Netherlands
The Boulevard, Langford Lane, Kidlington, Oxford OX5 1GB, United Kingdom
50 Hampshire Street, 5th Floor, Cambridge, MA 02139, United States

Notices
Knowledge and best practice in this field are constantly changing. As new research and experience
broaden our understanding, changes in research methods, professional practices, or medical
treatment may become necessary.

Practitioners and researchers must always rely on their own experience and knowledge in evaluating
and using any information, methods, compounds, or experiments described herein. In using such
information or methods they should be mindful of their own safety and the safety of others, including
parties for whom they have a professional responsibility.

To the fullest extent of the law, neither the Publisher nor the authors, contributors, or editors, assume
any liability for any injury and/or damage to persons or property as a matter of products liability,
negligence or otherwise, or from any use or operation of any methods, products, instructions, or ideas
contained in the material herein.

ISBN: 978-0-323-99503-0

For information on all Elsevier publications
visit our website at https://www.elsevier.com/books-and-journals

Publisher: Joseph P. Hayton
Acquisitions Editor: Kathryn Eryilmaz
Editorial Project Manager: Ali Afzal-Khan
Production Project Manager: Omer Mukthar
Cover Designer: Mark Rogers

Typeset by STRAIVE, India

Working together
to grow libraries in
developing countries

www.elsevier.com • www.bookaid.org

Contents

Part One
Smart city framework and implementation

Part Four
Smart environment

Part Five
Smart transportation

Part Six
Tackling cyber attacks

Part Seven
Smart communications

Contributors

Ansari Jameel Ahmad
Department of Engineering Sciences, Faculty of Science and Technology, Vishwakarma University, Pune, India

Sunitha Purushottam Ashtikar
School of Business, SR University, Warangal, Telangana, India

Manjulata Badi
Department of Electrical and Electronics Engineering, Alliance University, Bangalore, Karnataka, India

Bishwajit Dey
Department of Electrical and Electronics Engineering, GIET University, Gunupur, Odisha, India

Subhashini Durai
GRD Institute of Management, Dr. G.R. Damodaran College of Science, Coimbatore, Tamil Nadu, India

Rakesh Kumar Dutta
Department of Civil Engineering, National Institute of Technology, Hamirpur, Himachal Pradesh, India

Ahmed M. Ebid
Department of Structure Engineering & Construction Management, Faculty of Engineering, Future University in Egypt, New Cairo, Egypt

John S. Effiong
Department of Civil Engineering, Michael Okpara University of Agriculture, Umudike, Nigeria

O.V. Gnana Swathika
Centre for Smart Grid Technologies, School of Electrical Engineering, Vellore Institute of Technology, Chennai, India

Tammineni Gnananandarao
Department of Civil Engineering, Aditya College of Engineering and Technology, Surampalem, Andhra Pradesh, India

Varsha Himthani
Manipal University Jaipur, Jaipur, Rajasthan, India

T. Jesudas
Department of Mechatronics, Mahendra Engineering College, Namakkal, Tamil Nadu, India

Ginnes K. John
Department of Electrical and Electronics Engineering, Amrita School of Engineering, Amrita Vishwa Vidyapeetham, Coimbatore, India

Radha Karmarkar
India Smart Cities Fellowship (ISCF), National Institute of Urban Affairs under the Ministry
of Housing and Urban Affairs, New Delhi, India

Vishwas Nandkishor Khatri
Civil Engineering Department, IIT Dhanbad, Jharkand, India

Maneesh Kumar
Department of Electrical Engineering, Yeshwantrao Chavan College of Engineering,
Nagpur, India

Praveen Kumar
Department of Electronics and Communication Engineering, VIT-AP University,
Vijayawada, India

Sheila Mahapatra
Department of Electrical and Electronics Engineering, Alliance University, Bangalore,
Karnataka, India

S.K. Manju Bargavi
School of Computer science and IT, Jain (Deemed to be University), Bangalore, Karnataka,
India

Geetha Manoharan
School of Business, SR University, Warangal, Telangana, India

K.S.R. Murthy
Adhoc Faculty, Department of Electrical Engineering, National Institute of Technology,
Tadepalligudem, Andhra Pradesh, India

Vivin R. Nair
India Smart Cities Fellowship (ISCF), National Institute of Urban Affairs under the Ministry
of Housing and Urban Affairs, New Delhi, India

T.N.P. Nambiar
Department of Electrical and Electronics Engineering, Amrita School of Engineering, Amrita
Vishwa Vidyapeetham, Coimbatore, India

M. Pandya Nayak
Department of Commerce, PG Centre, Palamuru University, Mahabubnagar, Telangana,
India

Kennedy C. Onyelowe
Department of Civil and Mechanical Engineering, Kampala International University,
Kampala, Uganda; Department of Civil Engineering, Michael Okpara University of
Agriculture, Umudike, Nigeria; Department of Civil Engineering, University of the
Peloponnese, Patras, Greece

Aadyasha Patel
School of Electrical Engineering, Vellore Institute of Technology, Chennai, India

Vivek Prakash
Department of Electrical and Electronics Engineering, Banasthali Vidyapith, School of
Automation, Tonk, India

Deepak Prashar
School of Computer Science and Engineering, Lovely Professional University, Phagwara, India

Saurav Raj
Department of Electrical Engineering, Institute of Chemical Technology, Jalna, India

Gunaseelan Alex Rajesh
Sri Venkateswara Institute of Information Technology and Management, Coimbatore, Tamil Nadu, India

P. Srinath Rajesh
Department of EEE, Abdul Kalam Institute of Technological Sciences, Kothagudem, Telangana, India

Aman Singh Rajput
Urban Infrastructure and Tourism, IPE Global Limited; India Smart Cities Mission, New Delhi, India

Col B.S. Rao
School of Business, SR University, Warangal, Telangana, India

K. Srinivas Rao
Department of Electrical and Electronics Engineering, GIET University, Gunupur, Odisha, India

Mamoon Rashid
Department of Computer Engineering, Faculty of Science and Technology, Vishwakarma University, Pune, India

Abdul Razak
School of Business, SR University, Warangal, Telangana, India

V. Sathiyamoorthi
Department of Computer Science and Engineering, Sona College of Technology, Salem, India

G.R.K.D. Satya Prasad
Department of Electrical and Electronics Engineering, GIET University, Gunupur, Odisha, India

Vigya Saxena
Indian Institute of Technology IIT (ISM), Dhanbad, Jharkhand, India

Sachidananda Sen
Department of Electrical and Electronics Engineering, SR University, Warangal, Telangana, India

M. Senbagavalli
Department of Information Technology, Alliance University, Bangalore, Karnataka, India

Swetha Shekarappa G.
Department of Electrical and Electronics Engineering, Alliance University, Bangalore, Karnataka, India

M.R. Sindhu
Department of Electrical and Electronics Engineering, Amrita School of Engineering, Amrita Vishwa Vidyapeetham, Coimbatore, India

Tapas Ch. Singh
Department of Electrical and Electronics Engineering, GIET University, Gunupur, Odisha, India

P. Tejaswi
School of Electrical Engineering, Vellore Institute of Technology, Chennai, India

B. Vikram Anand
Department of Electrical and Electronics Engineering, GIET University, Gunupur, Odisha, India

Dharmendra Yadeo
Department of Electrical and Electronics Engineering, SR University, Warangal, Telangana, India

A study on the perceptions of officials on their duties and responsibilities at various levels of the organizational structure in order to accomplish artificial intelligence-based smart city implementation

Geetha Manoharan[a], Subhashini Durai[b], Gunaseelan Alex Rajesh[c], Abdul Razak[a], Col B.S. Rao[a], and Sunitha Purushottam Ashtikar[a]

[a]School of Business, SR University, Warangal, Telangana, India
[b]GRD Institute of Management, Dr. G.R. Damodaran College of Science, Coimbatore, Tamil Nadu, India
[c]Sri Venkateswara Institute of Information Technology and Management, Coimbatore, Tamil Nadu, India

1. Introduction

1.1 Smart Cities Mission (SCM)

State/Union Territory and Urban Local Bodies (ULBs) will be the promoters and shareholders of a limited liability company (SPV) incorporated in accordance with the Companies Act, 2013, in which they will hold a 50/50 equity stake. The SPV will be required to adhere to all regulatory and monitoring mechanisms outlined in the Companies Act. Most of the selected smart cities have the same organizational structure as proposed in their Smart City Proposal [1]. A typical hierarchical organization structure has been adopted across most of the SPVs. It is important for a project implementing agency to organize itself in line with the needs of the projects and subproject components to be designed, constructed, and implemented. The preferred organization model for such specialized agencies is a flat organization with multiple domain expert groups that are focused on implementation of the specific projects. The SPVs as implanting agencies need to adopt a more project focused organization structure. The success of the project depends on the effective functions of the stakeholders after finding the solutions for the challenges that are faced during the implementation.

Artificial Intelligence and Machine Learning in Smart City Planning
https://doi.org/10.1016/B978-0-323-99503-0.00007-7

2. Smart city assessment

Smart city assessment is a "nascent field with enormous potential for future development." Smart city implementations should be evaluated, so that their strengths and weaknesses can be documented; future improvements can be planned and stakeholders can be informed about how well various target goals have been met. SCA tools can be used to present city rankings, highlighting the best and worst locations for specific activities, which is an important tool for assessing the attractiveness of urban regions [2]. It is also possible to present city rankings, emphasizing the best and worst locations for various activities, using SCA software tools. With the SCA and its resulting rankings, cities can gain an advantage over their competitors. A city's international image benefits greatly from a highly regarded city ranking, and this can play a significant role in the marketing strategy of that city [3]. Some of the advantages of a city's high rank.

1. Cities' rankings draw attention to important issues that affect citizens' quality of life, and they encourage healthy competition among municipalities.
2. Cities are ranked based on their ability to foster economic growth in their respective regions.
3. The decisions of regional actors must be made in a way that is understandable to the general public.
4. Positive changes are being observed throughout the country, not just in the region.
5. The findings may be used to stimulate the learning effects of local actors if they are thoroughly examined.
6. Disadvantages are included to using different rankings for cities.
7. The complex interrelationships that exist in regional development are frequently overlooked in city rankings.
8. Long-term development plans may be jeopardized because the discussion is so narrowly focused on the bare rank.
9. Preexisting stereotypes may be reinforced.

3. Challenges in SCM

There is a critical capacity gaps on multiple levels both horizontally and vertically. Due to the leap fogging nature of the project both on the transformational plane as well as the technological plane the solution to the capacity, building challenge is not an easy one. It is important that the proposed Capacity Building Framework for the Smart Cities Mission, especially for the Smart Solutions Component must be carefully designed to assimilate the process of innovation but also be aware of the indigenous as well as international challenges of smart cities. The Framework cannot be borrowed as it is from global smart city experiences and solutions, and it must be designed to meet the local challenges and ground realities and the Smart Solutions, and approaches will have to be localized and adapted to the specific requirements of the Indian smart cities. The key gaps have to be address through building adequate human resource both internal and external and would also require continuous skill enhancement. The capacity building challenges can be summarized in to six main categories as listed below (Fig. 1):

(a) Governance, strategy, and institutional challenges

The smart solution planned under the program would require leadership and governance capabilities to achieve cooperation and collaboration among multiple city agencies. To ensure departmental coordination and alignment, overcome inefficiencies within departments, attract qualified ICT professionals and relevant Smart Technology Experts, and have qualified in-house project management experts who can manage technology application projects, the Smart City SPV project team would require assistance. Apart from this, a rational

Fig. 1 Challenges in smart city management.

distribution of inputs and revenues across SPV and ULB must also be addressed for a win–win on both sides for a smart and inclusive growth of the city.

(b) Technical challenges

The smart city teams are facing a large set of technical problems and issues. Several examples include the deployment of integrated city infrastructure and service platforms, the resolution of IoT/machine-to-machine communication issues, the definition and adoption of interoperability standards, particularly for the ICCC, the provision of analytical methods necessary to integrate multiple types of data from numerous sources, the optimal use of data sharing and integration for improving city operations efficiency.

(c) Service delivery challenges

The smart cities are under pressure to provide services to rising urban population in their respective cities. For example, increased demand for power, energy, water, and sanitation; increased pressure on housing and transportation systems; increased public safety by reducing crime and emergency response times; reducing traffic congestion; ensuring the construction of comfortable city facilities and buildings and improving service quality by providing innovative services and streamlining and tailoring services to address city issues.

(d) Financial and procurement challenges

The initial funding has been made available; however, the smart city program is not a grant-based projects modeled on the earlier pattern. The Smart Cities Mission envisages a paradigm shift to a partially granted but self-funded and managed project. Decisions to ensure longer-term financial sustainability, technical and operational feasibility, the use of innovative procurement models and approaches, including PPP and other partnership models, risk and financial resource sharing, addressing a potential lack of investor capacity, ensuring the construction of cost-effective buildings and facilities, lowering operational costs, and ensuring the long-term sustainability of the delivered solutions. To execute the smart city successfully, new financial and procurement instruments would be required increasingly to bridge the gap between grant and project requirement. Some such practices are already being used like innovative procurement approaches such as SWISS Challenge, and Reverse Auction.

(e) Project management challenges

The city government and SPV are new to these challenging complex and technical projects. Innovative governance and project management approaches and models, such as involving the private sector in piloting, testing, and operating new digital solutions, participatory governance through the adoption of citizen decisions and proposals, defining the appropriate role for private sector actor interventions—defining where, when, and how they should be engaged, attracting talent, and managing the expectations and participation of diverse stakeholders, are all supported by centrist institutions. The cities are implementing the smart solutions across the whole city but some of the solutions are location based that requires specific coordination with the line departments and other parastatal bodies. A comprehensive data management and knowledge architecture would also be needed for a smooth coordination across cities and state.

(f) **Project implementation challenges**

The major challenges that the SPV face in implementation of the projects are limited human resources or lack of staff and lack of capacities or limited skill set with the available staff. From digital solutions perspective, these challenges go way beyond digitization projects and would involve innovative solutions and experts who understand both the domain sector and also ICT tools and solutions that can be used to address the challenges. For the study team, the identification of the challenges is a big challenge due to the complexity and the scale of the transformation being attempted under the Smart Cities Mission. This project is a giant leap from a sectoral-driven city governance approach to an integrated smart city with use of cutting-edge technology. This requires a targeted capacity building, skill building and a continuous improvement and upgrade program, which will also need to be modified based on feedback loop for corrective measures.

4. Stakeholders involved in SCM

4.1 Involvement of stakeholders in smart cities

The Organization for Economic Cooperation and Development (OECD) defines a smart city as one that has a high level of stakeholder engagement as an input into the process of shaping a smart city (OECD) [4]. A smart city has a number of important stakeholders, including the following: city residents, nongovernmental organizations, and institutions of higher learning like universities are all factors to be taken into account when looking at the city's future (firms and entrepreneurs) (Fig. 2).

Smart city development necessitates the incorporation of stakeholders and partnerships that increase civic participation and leverage the role of

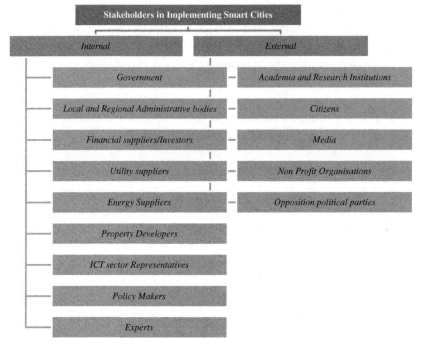

Fig. 2 Stakeholders involved in implementing smart cities.

the private sector in local decision-making. All of these imply a fair and equitable distribution of power among stakeholders, whether it is through simple communication and involvement or through full coproduction, delivery, and evaluation. As a result of digital innovation and technology, a wider range of urban residents and other stakeholders can participate in policy development and implementation in a more collaborative manner [5]. An evaluation of stakeholder engagement has a number of advantages, including the following:

- Increase the accountability of decision makers by determining whether public and institutional resources, along with stakeholder efforts, are being used appropriately and efficiently.
- It was our job to help determine whether the engagement process was a success and compile a list of lessons learned that could be used in future processes.
- Take part in identifying and managing potential threats.
- At the beginning of a process, map out the various viewpoints held by various stakeholders and identify potential roadblocks.

5. Duties and responsibilities of the officials in executing the AI

Smart city deployments are described in great detail in a variety of ways, but there is no real structure for understanding or replicating them or for disseminating the information that has been gathered about them. There are no relevant results when you conduct a web search solely on the basis of a set of "keywords." Inconsistent and unclear information presentation in a particular problem, because of this, it is getting more difficult to spread successful policies and practices to other cities. As a result of the lack of a standardized method for reporting successful smart city solutions or case studies, there is a great deal of wasted time and effort [6]. As an example, environmental sustainability benchmarking and common standards have advanced significantly over the last decade in botany, which has classification systems in place for more than a century. In order to identify and reuse data on stakeholder roles, policy requirements, and business models associated with smart city initiatives, cities require a structured and well-defined template for best practices and policies. When developing a smart city plan, it is important to identify the stakeholders who have the most impact on the city's initiatives and operations (Fig. 3).

Fig. 3 Beneficiaries for stakeholders.

6. Importance of roles of stakeholders in implementing SCM

Local governments and municipal governments
- Monitoring of the city's performance is done in order to improve its international reputation and competitive position in the eyes of investors, creative residents, and the general public.
- To demonstrate the value of smart city investments and interventions, cite examples from the real world.
- Smart city planning requires an understanding of a city's assets and liabilities.
- The city's overall position in its effort to become smarter should be kept in mind as well as the progress made toward predetermined goals and targets.
- Recognize the socioeconomic and environmental consequences of initiatives to develop smart cities.
- Be familiar with the technological specifications that are required for smart city projects.
- Utilize peer experiences (when assessment involves benchmarking).
- Identify and display examples of best practices so that others can benefit from them.
- Increase the level of transparency in governance.
- Encourage dialogue among a diverse range of stakeholders, which may result in a more effective use of available resources.

Institutions of finance and investors
- Evaluation of completed or ongoing projects using evidence-based methodology.
- Methods for prioritizing funding allocation that are based on scientific principles.
- Increased capacity for determining the most suitable locations for future investment.
- It is essential to be able to recognize and take advantage of new opportunities in the marketplace.
- Members of the investigative or research community.
- Improve the performance of smart cities with new ideas.
- The smart city concept should be simplified to its simplest form.

Nationals/citizens

- There was an increase in public awareness of the benefits of smart city initiatives.
- Investing in the future requires the ability to make sound judgments.
- Promoting smart city development activities and encouraging citizens to voice their preferences for the city's future.

7. Conclusion

The challenges are to be addressed and appropriate measures are to be taken to close the gaps by which the Smart Cities Mission (SCM) will be more successful. The role of stakeholders is critical to the success of a successful SCM, and it is believed that three major goals can be achieved through collaboration. First and foremost, the hierarchy of physical city components can be used to develop a taxonomy or typology that will allow cities to compare and contrast relevant content in their respective jurisdictions. Then, through the use of stakeholder roles, it assists in defining who is accountable for what is done [4]. The fact that this component is frequently overlooked in city discussions can be attributed to a lack of knowledge about how to implement smart city solutions in a variety of settings. Furthermore, the content catalogue of a city is straightforward to navigate. In addition, many other things will be possible as a result of these outcomes. AI is important to identify areas where ICT solutions can be implemented in urban areas and how they can be implemented there [7]. By developing policy guidelines for the government to follow, it also encourages private sector participation in urban development initiatives. It assists in performing a city gap analysis, which is necessary for cities to be able to compare themselves fairly to other cities, as well as in developing a structured case study template, which is necessary for assembling a variety of business models for similar smart city initiatives.

References

[1] Cities, Smart, and Inclusive Growth, Measuring Smart Cities' Table of Contents, 2020 (December).
[2] C. Patrao, P. Moura, A.T. de Almeida, Review of smart city assessment tools, Smart Cities (2020) 1117–1132.
[3] European Commission, 100 Climate-Neutral Cities by 2030—By and for the Citizens, Interim Report of the Mission Board for Climate-Neutral and Smart Cities, EC, Brussels, Belgium, 2020.

[4] OECD, Smart Cities and Inclusive Growth. Building on the Outcomes of the 1st OECD Roundtable on Smart Cities and Inclusive Growth, OCED, Paris, France, 2020, pp. 1–59. Available online: https://www.oecd.org/cfe/cities/OECD_Policy_Paper_Smart_Cities_and_Inclusive_Growth.pdf.

[5] European Commission, 2020. https://ec.europa.eu/knowledge4policy/foresight/topic/continuing-urbanisation/developments-and-forecasts-on-continuing-urbanisation_en.

[6] S. Bannerjee, J. Bone, Y. Finger, C. Haley, 2016. https://digitalcityindex.eu/uploads/2016%20EDCi%20Construction%20Methodology%20FINAL.pdf.

[7] Ministry of Housing & Urban Affairs (MoHUA), Integrated Command and Control Center Maturity Assessment Framework and Toolkit Maturity Assessment Framework and Toolkit to Unlock the Potential of Integrated Command and Control Centers (ICCCs), 2018 (December).

PART ONE

Smart city framework and implementation

CHAPTER TWO

Integration of IoT with big data analytics for the development of smart society

Mamoon Rashid[a], Ansari Jameel Ahmad[b], and Deepak Prashar[c]
[a]Department of Computer Engineering, Faculty of Science and Technology, Vishwakarma University, Pune, India
[b]Department of Engineering Sciences, Faculty of Science and Technology, Vishwakarma University, Pune, India
[c]School of Computer Science and Engineering, Lovely Professional University, Phagwara, India

1. Introduction

The Internet of Things (IoT) term relates to the billions of smart devices worldwide that are now Internet-connected, many of which gather and exchange data. Physical objects such as refrigerators, vehicles, or any electrical and electronic gadgets also act as the part of IoT systems. Sensors and computer-based chips are used in IoT to collect large amount of data for those devices that cannot be connected to the Internet [1,2]. Whenever these collected data from smart devices show enormous volume, velocity, and variety, then the importance of data tends to come together with IoT comes into the picture. The data collected through various sensors is voluminous and holds information in form of both unstructured and structured ways. The velocity challenge of big data is the speed at which it is analyzed, and it also demonstrates its appearance in terms of IoT-based data. Variety is the data that appears in different forms and is one of the important sources of IoT data. The main challenge in IoT is how to manage huge amounts of data that IoT devices produce simultaneously. Big data analytics with their continuous streaming nature of the information has the capabilities to accommodate these IoT-based data. IoT connects with big data, and it is clear that the two phenomena match each other. IoT-based data that are gathered by the devices must be filtered so that they are relevant and useful. Big data analytics tools are used to extract information from raw data to obtain useful knowledge in IoT applications to add smartness to it. The size of data conducted in IoT is radically different, however, and analytics systems will take care of specific approach es to collect reliable results. The

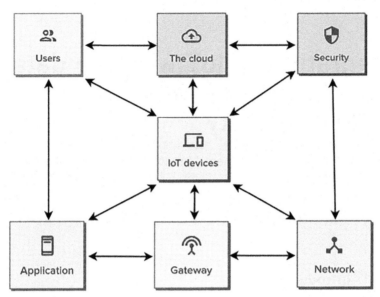

Fig. 1 Broader view of IoT ecosystem.

relation between IoT and big data is outlined by Ahmed et al. [3], Singh et al. [4], and Samkria et al. [5]. The IoT system sketch for IoT infrastructure is shown in Fig. 1.

Some of the major application domains like healthcare, shipping, agriculture, and logistics organizations used big data and IoT together to provide additional insight and assessment. In agriculture domain, the crop fields are directly linked to monitoring equipment for observing levels of moisture in fields and this information is subsequently given for timely information to farmers. The shipping organizations use large amount of sensor data and big data analytics to maximize the productivity of delivering different vehicles to preserve their fuel economy and speed.

This chapter discusses the background of IoT and big data along with terminologies in Section 2. Various standards and protocols are studied in Section 3. Interrelation of data analytics with IoT is discussed in Section 4. Section 5 provides the novel outline of big data analytics architecture for IoT applications keeping various requirements into consideration. Section 6 provides challenges and open issues in IoT and data analytics. Section 7 provides conclusions and future directions.

2. Key terminology of IoT and big data

IoT is the connectivity of different physical systems for information exchange via the Internet. Numerous studies have systematically described

IoT from various perspectives. In research [6], the authors have characterized IoT based on three A's—anywhere, anytime, and any media leading in a balance among both man and radio in the ratio of one: one. The methods to ensure data protection in public cloud storage are addressed in Rashid and Chawla [7] and Akram et al. [8] by offering an expanded role-based access control framework where authorized clients can download information only in terms of tasks with defined permissions and limit the access of unauthorized users by introducing variable constraints. IoT can be defined by Van Kranenburg [9] as an interoperating protocol-based global network where the physical and virtual objects have characteristics and attributes with obviously self-managing abilities. In article Atzori et al. [10], the authors described IoT as an innovative paradigm of communication systems where objects like cell phones, wireless sensors, RFIDs, and actuators communicate with one another and their surroundings in order to achieve the desired common objectives. Smart home systems are suggested by Singh and Rashid [11] to track and manage devices by combining cloud storage network services with IoT. Cloud computing enables remote access for smart devices, and IoT revolutionizes normal devices as intelligent devices [12]. Figure 2 displays the design of an IoT model.

The IoT is defined as an internetworking of smart and physical devices where software, sensors, and actuators are embedded to enable machines to interact with each other [13]. In connecting devices, the concept of IoT is noticed when communication takes place with customer service systems, digital marketing, and data analysis applications. IoT's scenario-based architecture is provided by Boyes et al. [14]. The different approaches are

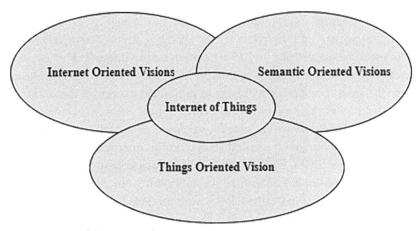

Fig. 2 Internet of Things paradigm.

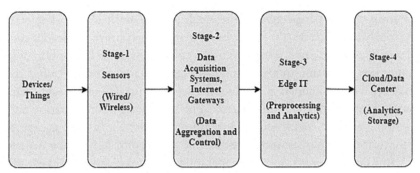

Fig. 3 Stage-wise architecture of IoT.

inputted into the architectural stage, where the IoT sensors are either wired or wireless. In the next stage, the Internet gateways and data collection systems are used for the aggregation and control of data. In the next phase, preprocessing and different analytics are conducted for which data center or cloud services are used. The entire process is visualized in Fig. 3, in the IoT architecture.

2.1 Need of IoT

IoT is one of the most popular technologies, which is having its future use. According to Gartner [15], IoT is going to cover 26 billion units by 2020 and will single handily control the supply chain operations and the various information that is required by these operators. For all real-time applications and environments, IoT is converting business processes into products with more accuracy and visibility. The importance of IoT was revealed by the UK government in its budget of 2015 where the next information revolution was referenced in terms of interconnectivity from home appliances to transport devices. A finance of £40,000,000 was allocated for IoT research for its future use. The biggest plus of IoT is its reach from a distance. For instance, one can trigger all home-based appliances by making smart IoT-based home where the sensors or signals will help an individual to control these gadgets over the Internet. Use of IoT is there in verticals of manufacturing and health care as well where the remote assistance of quality medical practitioners is given from urban to rural areas over the Internet. The IoT makes traditional systems open-ended in the form of multiple applications-multiple devices from close-ended systems of one application-one device. The open-ended approach in IoT systems makes it interoperable, which allows its integration at the horizontal level [16]. The same is depicted in Fig. 4.

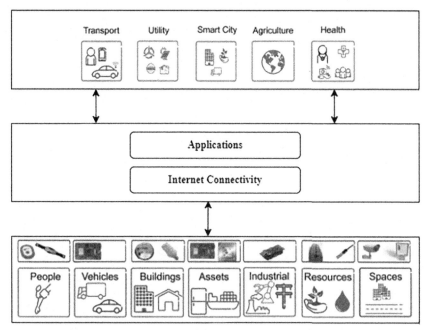

Fig. 4 Internet of Things usage in real life.

2.2 Use of big data for faster computations

Big data are a concept for huge information collections. Content is produced from different technology outlets in the modern world that have contributed to deep learning development. Significant information has often been called a movement that changes our way of living, working, and thinking [17]. The technological invention is framing so that information is released and can be composed smarter to use broad bulk statements [18]. Data analytics is in petabytes and is further distributed in standardized, conventional, and population standard deviation. It is known collectively between 3Vs and 4Vs. The amount, pace, and variance apply to 3Vs. The process contributes to the enormous testimony produced once a morning, while the step is equalized inflation and quite fast; the results are obtained for evaluation. A collection of document types include variable data mostly of unstructured types. The fourth point concerns accessibility and transparency for truthfulness [19]. Predictive analysis is described in Fig. 5.

This same capability to derive quality through big data relies on analysis; assessments are the cornerstone of the information transformation [20]. The implementation of big data includes the compilation, alteration, and continuation of the collection of datasets, so it is easy for market analysts to obtain it

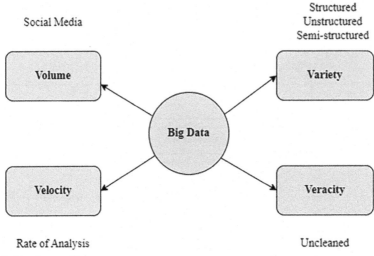

Fig. 5 Big data outline summary.

can ultimately provide consumers with information services. The essence of predictive optimization is its translation into a frequent model for organizations of vast amounts containing disordered empirical knowledge [21]. The purpose of big data management is to focus on making informed judgments for societies. Humans cannot comprehend a future lacking of a database setup, a location where any component concerning an entity is forgotten immediately during usage, every procedure conducted, or all aspects that can be registered. It causes a shortage of capacity in collecting useful expertise and in-depth research offering great possibilities. The contact information of the franchisee, products, acts done, and business activities are necessary throughout the daily time. Statistical information is the prime segment that any business relies endlessly on. Big data are comparable to a tiny CPU yet larger in volume and larger in statistics, so various methods such as strategies, software, and frameworks are required. The focus is on finding solutions to existing issues and underlying arguments. It consists of compiling comprehensive and complex resources, which are complicated by the use of conventional system integration techniques. The size of big data is rising gradually.

3. Standards and protocols of IoT

According to researchers, it seems that there is no clear line for the categorization of IoT standards, but the main standard protocols used are based on IoT data link protocols [22]. The majority of physical and MAC layer protocols have been used according to different IoT specifications.

IEEE 802.15.4: This is a MAC layer-based standard, and it specifies headers, source and destination addresses, format of frames, and identification in communication between the nodes. In IoT, relatively cheap communication and high efficiency are facilitated by the use of the channel hopping and time synchronization.

IEEE 802.11ah: In conventional networking, this type of standard is used as IEEE 802.11 (Wi-Fi) for different IoT applications. This standard can be used for sensor-friendly power communication and supports lower overhead. This standard includes synchronization frame functionality, smaller MAC frames, and powerful bidirectional packet exchange.

Wireless HART: Wireless HART relies on the MAC layer and uses time division multiple access scheme. For encryption, various algorithms used by this standard make it more efficient and secure as compared to other standards.

Z-Wave: Z-Wave works essentially on the MAC layer and was specifically designed for home automation. Master-Slave configuration is used by this standard where the small texts are sent by the master to the slaves, which are used for point-to-point short distances.

IEEE 802.15.4 (ZigBee): ZigBee is used for low-powered device that is used in personal area network communication. In applications like healthcare systems and remote controls, ZigBee provides reliability and accuracy. This standard is intended for communications to medium level.

DASH7: This standard is basically wireless sensor protocol used on the MAC layer, and it is typically used in RFID-enabled devices. It works on master-slave architecture, and it is particularly suited for IoT applications such as Industry 4.0, and it is used for IPv6 addressing.

Home plug: This standard is based on MAC protocol and is largely used in smart energy and smart grid applications. The elegance of this model lies in its power-saving mode, whereby nodes can take long nap when not necessary and only awake when needed.

4. Data analytics for IoT

IoT and data are interrelated with each other. The data collected by IoT devices result in values only when such data are analyzed in data analytics platform. Data analytics is basically a kind of process, which analyzes larger datasets with varying characteristics and results in meaningful insights.

4.1 Characteristics of IoT-generated data

The characteristics of IoT-based data usually come under four dimensions: accuracy, consistency, completeness, and timeliness.

Accuracy: The accuracy in IoT network is usually calculated in terms of value differing from each device in a network. As an example, if there are 20 devices in IoT network, which are having task to report temperature, then the accuracy in this case will monitor the reporting of temperature from all devices in terms of their deviations from each other.

Consistency: The values generated in IoT network from all the devices linked to it are checked for their consistency while logged in big data environment. For example, if values related to multiple events are reported by IoT network on mobile devices with geolocation, then consistency is checked for values to be same or different.

Completeness: This characteristic makes sure that values have been accumulated from IoT network without any gaps or missing values.

Timeliness: Usually, data are having streaming nature and come from multiple devices in IoT network, so synchronization is always required. Timeliness characteristic makes sure that the data values are collected within acceptable time frame.

4.2 Big data analytics life cycle

The structure of huge information comprises many different stages, illustrated in the following steps:

Data acquisition: The first step of the analytics life cycle is to obtain and collect data. With rising data sources, data generation is increasing at a rapid rate. The data processing stage of the life cycle of big data analytics determines the type of frameworks that may demand data processing.

Data extraction: In the data acquisition process, the data produced and retrieved are not in a functional form. There remains a great deal of complexity and unwanted information. Filtering information in the appropriate contexts and integrating data from different variations into one universal platform is the primary factor in the data extraction process.

Data collation: Data collection is one of the main stages of the phase of big data analytics, in which data from various sources are operating for analytical purposes. Different sources of information often give a clearer image in the process of research. For instance, data from many sources are in climate modeling structures that reveal the regular moisture, temperature, humidity, etc.

Data structuring: This process has to arrange the information systematically. The data in a standard format are easy to view and manipulate. Information structuring organizes the records properly. However, new systems are emerging that allow the processing of unstructured data feasible as well.

Data visualization: In this phase, the data of the comprehensive data monitoring process are exhibited in the determined layout. Mapping of data encompasses regions and allows graphic and pattern perspectives. Source code could even be employed to collect information or analyze trends, so it is useful to civilize the results.

Data interpretation: Data analysis is the last step in the continuum of big statistics whereby precious evidence through content processing is collected. The gathered knowledge has various categories: Observational examination requires an interpretation of current occurrences and behavior. Forward assessment involves evaluation developments and variable selection previously established.

4.3 Types of data analytics technologies for IoT

Different types of data analytics are used in IoT to draw insights out of collected data.

However, the most common ones are listed below:

Streaming analytics: In this type of analytics, we usually deal with the data in motion, which needs processing in real time. Data streams that are real time in nature are analyzed for the detection of critical situations. This streaming is quite useful in traffic analysis and stock transaction-based IoT applications.

Spatial analytics: This kind of data analytics is used for analyzing data in objects, which are having spatial relation in terms of geographical patterns. Spatial analytics is useful in lo-cation-based IoT applications like IoT-based parking applications.

Time series analytics: This kind of data analytics is used for analyzing data, which is time-based in nature. Time series analytics helps in revealing the patterns in IoT applications. This kind of analytics is useful for healthcare-based IoT applications and weather forecasting IoT applications.

5. IoT-based big data analytics platform

This section discusses the various requirements of big data analytics for IoT data analytics and proposes the novel outline of big data analytics pipeline for IoT applications. Next, the comparison of IoT-based platforms is done on the basis of flexibility, scalability, various services, ease of connections, and pricing.

5.1 Requirements

In the last few years, the need for big data systems and the IoT analytics ecosystem has increased several times and offers useful advantages and improvements in decision-making procedures. Therefore, for better analytics in its data processing, the criteria and demands of big data analytics frameworks in IoT have expanded. The incorporation of big data pipelines in the IoT has altered the way information is processed and analyzed. Data analytics for the extraction of useful perspectives will efficiently process the larger quantities of data produced by different sensor devices. This chapter outlines the key criteria needed by the IoT ecosystem for big data analytics platform data processing.

Connectivity: Broadband connections are the significant criteria for big data analytics on large volumes of device anomaly detection in the IoT setting [23]. Accurate connectivity is a way of linking high-performance infrastructures to IoT services with multiple types of artifacts.

Streaming analytics: This kind of data analytics is another primary feature in IoT environments and contracts with real-time data transmission. For the identification of critical conditions, information sources that are real time in the design are analyzed. Big data analytics need and manage information of streaming data in various real-time systems [24].

Storage: The processing of large amounts of data generated by different IoT artifacts on assured utility storage units with low bandwidth variables in its coherent is another primary feature in IoT-based big data platforms. many other IoT networks, Machine to Machine (M2M) sensor nodes are commonly used to manage massive streams of sensor data and its distributed processing [25].

Quality of services: The standard of services on mobile devices and IoT detectors is related to resource utilization and is the prerequisite of IoT-based big data platforms. In an IoT-based channel where powerful data processing is demanding from different devices and objects to big data platforms, the quality of service must be quite effectual [26].

5.2 Proposed IoT-based big data analytics pipeline

Based on the requirements discussed, the authors are proposing a big data pipeline applicable for IoT environment for streaming data of smart cities by combining IoT architecture with big data analytics environment. The idea is to connect all the IoT sensor devices through wireless networks in the form of Wi-Fi, RFID, Bluetooth, and ZigBee. IoT gateway is used for making the communication with the Internet to use storage of cloud platform infrastructure. Cloud storage is used for storing huge amounts of data received from sensor devices. Later, these data are accessed and processed in big data analytics pipeline where Apache Spark is used for processing these real-time data. The set of APIs of cloud integration are available in Apache Spark, which will help to process real-time data on big data analytics platform. The overall architecture of IoT-based big data analytics pipeline is shown in Fig. 6.

5.3 Comparison with existing platforms

The comparison of IoT-based platforms is done on the basis of flexibility, scalability, various services, ease of connections, and pricing [27]. The comparison of different IoT platforms like AWS, Microsoft Azure, and IBM Watson is given in Table 1.

6. Challenges and issues of IoT and data analytics

Big data analytics solutions based on IoT are very helpful in overcoming many issues, but there are a range of challenges that need to be tackled.

Management of big data: Data produced by IoT wearable sensors are typically of a heterogeneous nature. Processing and storing that massive volume of data is a challenging task. With the available amount of network bandwidth, the response time and processing efficacy of IoT sensor devices become the core strategy.

Data stream mining: In IoT applications, typically streaming of information is produced in a lot of volumes that are usually unstructured photographs and streaming video data. It is a challenging task to adhere data mining methods to such information for accurate analysis.

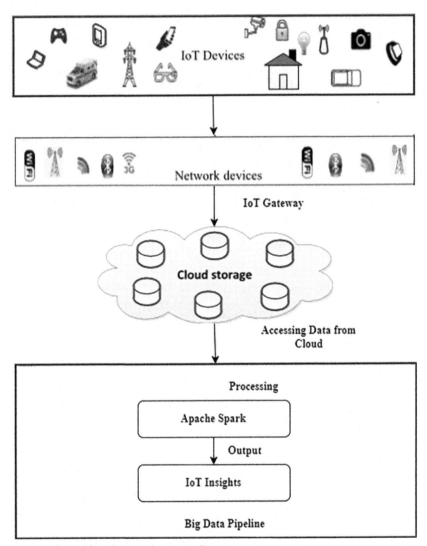

Fig. 6 IoT-based big data analytics pipeline.

Privacy concern: IoT devices have large quantities of data relating to healthcare, smart grids, and smart city developments where privacy issues are of paramount concern. Protection of privacy is the main issue that needs to be tackled to ensure improved operational efficiency.

Security risks: In IoT systems where devices are increasing in numbers, security risks are still present. Tackling different types of data theft and cyberattacks while trying to deal with the information received from IoT-based operating systems is a major challenge.

Table 1 Comparison of existing IoT platforms.

Features/IoT platform	AWS	Microsoft Azure	IBM Watson
Hardware usage	Intel, Texas Instruments, Broadcom	Intel, Texas Instruments, Raspberry Pi2	Raspberry Pi2, Texas Instruments, Arduino
IoT protocol followed	MQTT, HTTP	MQTT, HTTP	HTTP, MQTT
Coding language	NodeJS, C, Java	.Net, NodeJS, C	Java, Python, C, NodeJS
Prices	Pricing is for million messages	Pricing is for IoT hub with number of messages and devices	Pricing for devices, storage, and traffic

7. Conclusions and future directions

Data generated by various IoT-based sensor devices have increased many folds in the last few years. As a result, processing and analyzing these huge amounts of data in the interaction of IoT and big data analytics platforms are necessary. The massive rise in the number of devices linked to IoT and the exponential rise in the usage of data only illustrate how big data growth ideally overlaps with IoT growth. Big data analytics would also move at the edge for real-time decision-making with IoT. This chapter highlights the various challenges, which arise in the IoT environment, and proposes a pipeline based on big data analytics for processing the data stored in cloud storage infrastructure. This novel architecture would, hopefully, address many issues in IoT data processing, especially the processing of real-time streaming IoT-based data. Several challenges have also been emphasized; those need to be addressed in the future to enhance the reliability of IoT platforms.

References

[1] F.J. Riggins, S.F. Wamba, Research directions on the adoption, usage, and impact of the internet of things through the use of big data analytics, in: 2015 48th Hawaii International Conference on System Sciences, IEEE, 2015, January, pp. 1531–1540.

[2] R. Singh, M. Baz, C. Narayana, M. Rashid, A. Gehlot, S.V. Akram, A.S. AlGhamdi, Zigbee and long-range architecture based monitoring system for oil pipeline monitoring with the internet of things, Sustainability 13 (18) (2021) 10226.

[3] E. Ahmed, I. Yaqoob, I.A.T. Hashem, I. Khan, A.I.A. Ahmed, M. Imran, A.V. Vasilakos, The role of big data analytics in internet of things, Comput. Netw. 129 (2017) 459–471.

[4] R. Singh, M. Baz, A. Gehlot, M. Rashid, M. Khurana, S.V. Akram, A.S. AlGhamdi, Water quality monitoring and management of building water tank using industrial internet of things, Sustainability 13 (15) (2021) 8452.

[5] R. Samkria, M. Abd-Elnaby, R. Singh, A. Gehlot, M. Rashid, M.H. Aly, W. El-Shafai, Automatic PV grid fault detection system with IoT and LabVIEW as data logger, CMC-Comput. Mater. Continua 69 (2) (2021) 1709–1723.

[6] L. Srivastava, Pervasive, ambient, ubiquitous: the magic of radio, in: Europe—An Commission Conference "From RFID to the Internet of Things", Bruxelles, Belgium, 2006, March.

[7] M. Rashid, E.R. Chawla, Securing data storage by extending role-based access control, Int. J. Cloud Appl. Comput. (IJCAC) 3 (4) (2013) 28–37.

[8] S.V. Akram, R. Singh, M.A. AlZain, A. Gehlot, M. Rashid, O.S. Faragallah, D. Prashar, Performance analysis of IoT and long-range radio-based sensor node and gateway architecture for solid waste management, Sensors 21 (8) (2021) 2774.

[9] R. Van Kranenburg, The Internet of Things: A Critique of Ambient Technology and the all-Seeing Network of RFID, Institute of Network Cultures, 2008.

[10] L. Atzori, A. Iera, G. Morabito, The internet of things: a survey, Comput. Netw. 54 (15) (2010) 2787–2805.

[11] P. Singh, E. Rashid, Smart home automation deployment on third party cloud us- ing internet of things, J. Bioinform. Intell. Control 4 (1) (2015) 31–34.

[12] A. Jan, S.A. Parah, B.A. Malik, M. Rashid, Secure data transmission in IoTs based on CLoG edge detection, Future Gener. Comput. Syst. 121 (2021) 59–73.

[13] I. Lee, K. Lee, The internet of things (IoT): applications, investments, and challenges for enterprises, Bus. Horiz. 58 (4) (2015) 431–440.

[14] H. Boyes, B. Hallaq, J. Cunningham, T. Watson, The industrial internet of things (IIoT): an analysis framework, Comput. Ind. 101 (2018) 1–12.

[15] W. Gartner, Gartner Says the Internet of Things Will Transform the Data Center, 2014 (Retrieved 07 September 7).

[16] V. Tsiatsis, S. Karnouskos, J. Holler, D. Boyle, C. Mulligan, Internet of Things: Tech- nologies and Applications for a New Age of Intelligence, Academic Press, 2018.

[17] S. Madden, From databases to big data, IEEE Internet Comput. 16 (3) (2012) 4–6.

[18] D.E. O'Leary, Ethics for big data and analytics, IEEE Intell. Syst. 31 (4) (2016) 81–84.

[19] W.A. Günther, M.H.R. Mehrizi, M. Huysman, F. Feldberg, Debating big data: a lit- erature review on realizing value from big data, J. Strat. Inform. Syst. 26 (3) (2017) 191–209.

[20] H.V. Jagadish, J. Gehrke, A. Labrinidis, Y. Papakonstantinou, J.M. Patel, R. Ramakrishnan, C. Shahabi, Big data and its technical challenges, Commun. ACM 57 (7) (2014) 86–94.

[21] A. Gehlot, S.S. Alshamrani, R. Singh, M. Rashid, S.V. Akram, A.S. AlGhamdi, F.R. Albogamy, Internet of things and long-range-based smart lampposts for illuminating smart cities, Sustainability 13 (11) (2021) 6398.

[22] M. Rashid, H. Singh, V. Goyal, N. Ahmad, N. Mogla, Efficient big data-based storage and processing model in internet of things for improving accuracy fault detection in industrial processes, in: Security and Privacy Issues in Sensor Networks and IoT, IGI Global, 2020, pp. 215–230.

[23] E. Ahmed, M. Imran, M. Guizani, A. Rayes, J. Lloret, G. Han, W. Guibene, Enabling mobile and wireless technologies for smart cities, IEEE Commun. Mag. 55 (1) (2017) 74–75.

[24] M. Rashid, H. Singh, V. Goyal, Cloud storage privacy in health care systems based on IP and geo-location validation using K-mean clustering technique, Int. J. E-Health Med. Commun. (IJEHMC) 10 (4) (2019) 54–65.

[25] G. Suciu, V. Suciu, A. Martian, R. Craciunescu, A. Vulpe, I. Marcu, O. Fratu, Big data, internet of things and cloud convergence—an architecture for secure e-health applications, J. Med. Syst. 39 (11) (2015) 141.

[26] J. Jin, J. Gubbi, T. Luo, M. Palaniswami, Network architecture and QoS issues in the internet of things for a smart city, in: 2012 International Symposium on Communications and Information Technologies (ISCIT), IEEE, 2012, October, pp. 956–961.

[27] A. Botta, W. De Donato, V. Persico, A. Pescapé, Integration of cloud computing and internet of things: a survey, Future Gener. Comput. Syst. 56 (2016) 684–700.

CHAPTER THREE

Deep learning model for flood estimate and relief management system using hybrid algorithm

M. Senbagavalli[a], V. Sathiyamoorthi[b], S.K. Manju Bargavi[c], Swetha Shekarappa G.[d], and T. Jesudas[e]

[a]Department of Information Technology, Alliance University, Bangalore, Karnataka, India
[b]Department of Computer Science and Engineering, Sona College of Technology, Salem, India
[c]School of Computer science and IT, Jain (Deemed to be University), Bangalore, Karnataka, India
[d]Department of Electrical and Electronics Engineering, Alliance University, Bangalore, Karnataka, India
[e]Department of Mechatronics, Mahendra Engineering College, Namakkal, Tamil Nadu, India

1. Introduction

The main goal of this chapter is really to investigate the function of algorithms related to machine learning (ML), deep learning (DL), and artificial intelligence (AI) technology in smart city evolution. Today, some towns throughout the world are now adopting new technology and transforming into smart cities. The standard of living for citizens is improving as a result of new advancements. Any use of technologies, unfortunately, brings with it new concerns and obstacles. Green technology operations, such as intelligent atmosphere, intelligent citizens, and intelligent living, are linked and streamlined as personal and collaborative features within an integrated city operating system, based on the cumulative effect of device technology that supports smart city initiatives and a plethora of AI and their capabilities and implementations. This strategic sustainability permits cities to work considerably more efficiently while diminishing noise inside system operations. A single susceptible action by an individual or group in a smart city potentially puts the complete city in danger. The strategies described above are effective in determining the best policy for a variety of smart city related problems. Finally, we propose a number of research difficulties and prospective research directions in which the aforementioned methodologies might help actualize the smart city notion. We discuss in-depth descriptions of the uses of prior approaches in flood estimation and relief management systems in this chapter.

Artificial Intelligence and Machine Learning in Smart City Planning Copyright © 2023 Elsevier Inc.
https://doi.org/10.1016/B978-0-323-99503-0.00021-1 All rights reserved.

Flooding is a major geographical calamity that occurs frequently in some countries and infrequently in others. It is critical to remain vigilant and make early precautions to prevent unnecessary threats that endanger both person and property. Today, 50% of the world's inhabitants live in cities. As a result, persistent monitoring of urban characteristics is particularly useful for understanding people's living conditions, migration trends, and other issues. In terms of the repeat of acquiring practically all the authentic sensors, the city's change velocity is normally minimal. Natural disasters, on the other hand, can generate abrupt temporal disruption in the urban environment, necessitating a dependable and prompt reaction to the crisis situation. Furthermore, during a crisis, the acquisition atmosphere is predominantly caused, usually due to gravitational imbalances. If flooding happens, for example, clouds are likely to obscure an optical inspection of the afflicted area. A sensor that can gather information regardless of the weather or moment of the day is especially useful in this situation. The natural drainage system could be used for the management of urban flooding. Disregarding the natural drainage networks during urban development has caused flooding issues in the cities.

Drainage systems have failed as a result of increased quantity of water and the emission of harmful waste and other rubbish into sewage. Environmental anomalies are an inherent part of life, and they affect every component of the ecosystem. The importance of urban ecology in disaster risk mitigation is demonstrated by the mix of urban flood sources. The data were obtained and analyzed in relation to flood risk and urban planning. It highlights larger issues and offers several insights for managing floods in our country's urban and rural areas. The study will focus on the potential impact of flooding on the performance of urban transport and not the infrastructure damage. The study's goal is to identify regions where watercourses are inadequate for flood control. Failure of drainage systems will cause flooding in urban places and where there is no place for water outlet. To make flood mitigation more efficient and to implement nonstructural measures such as flood warning, flood proofing, flood forecasting, etc.

In this chapter, we have focused on deep learning-based model and Web application in which the affected drainage can be found and alert people who take necessary precautions based on it. Also, the Web application will provide information about the drain's width, height, and area of how much it has been affected so that we can assume when the flood is going to occur. The main aim of this chapter is to automate identification of the flood hot spots on roads and assess potential need for interventions in urban planning using machine learning technique, and its goal is to use geographical datasets to examine the effects of urban floods on traffic gridlock and to investigate the possibilities of employing hydrodynamic modeling to detect flood danger points. The other key objectives here would be establishing a correlation between traffic volume, traffic signal location with traffic congestions in urban road network during a flood event, and identification of the flood hot spots on roads and assessing potential need for interventions in urban planning. Vulnerability and contingency efforts are used to manage the current and future risk of floods. In the future, therapeutic interventions such as a conjunction of minimizing the time where feasible, regulating risk through conceptual or consensus sequence, monitoring risks through premiums, bracing, disaster management prepping, alarms, and corresponding risk to social gathering concerned could be implemented. A mobile application can be developed in the future to alert people with a message. This would be easy and helpful for all people residing in the area.

The natural drainage system could be used for the management of urban flooding. Disregarding the natural drainage networks during urban development has caused flooding issues in the cities. The purpose of this chapter is to find the areas where watercourses are incompetent for flood control. Failure of drainage systems will cause flooding in urban places and where there is no place for water outlet. The main aim of this chapter is to automate identification of the flood hot spots on roads and assess potential need for interventions in urban planning using machine learning technique. To make flood mitigation efficiency and implement nonstructural measures such as flood forecasting. In this chapter, we have proposed a Web application in which the affected drainage can be found and alert people who take necessary precautions based on that. Nowadays, half of the world's population lives in cities. As a result, constant monitoring of urban characteristics is particularly useful for understanding people's living conditions, migration trends, and other issues. With regard to the recurrence of acquisition of nearly all of the specific sensors, the town's change velocity is normally mild. Natural catastrophes, on the other hand, can generate abrupt temporal discontinuity inside the urban environment, necessitating a dependable and prompt reaction to the crisis situation. Furthermore, in the event of a crisis, the acquisition environment is frequently disrupted, typically by atmospheric phenomena. For example, if flooding occurs, it is quite likely that clouds will obstruct an optical investigation of the affected area. In this instance, a sensor that can collect data regardless of the weather or the time of the day is particularly desirable.

Flooding in metropolitan areas due to insufficient sewer systems has become a major concern. Because the value of houses and other structures has skyrocketed, the possible harm from protracted flooding may potentially reach the huge amounts of money. Citizens have to pay service charges and, as a result, demand their drainage systems to function properly without danger of collapse due to weather. Drainage systems built to withstand the most severe storm conditions, on the other hand, would be prohibitively expensive to install and maintain. The welfare of citizens and the preservation of their belongings must be balanced with technological and economic constraints when determining permissible flood frequencies, posing a difficult optimization problem.

Floods are the most common cause of destruction in cities. Improper construction of drainages with wrong measurements will lead to overflow of drainage. Most of the drainages are left unclosed and that will cause overflowing. In cases like that, it is better to make holes and attach small pipes where the extra water can be let out. In most hidden urban slum areas, the drainages affect the living of people who habitat there. Developing a Web application for checking whether the drainage is affected or not, we can prevent areas by taking necessary precautions. So the count will help to see how much drainage is affected as shown in Fig. 1.

The major goal of this research is to use geospatial datasets to examine the effects of urban floods on traffic jams and to investigate the feasibility of employing hydrodynamic modeling to detect flood danger points. The study will concentrate on the possible effects of flooding on urban transportation performance rather than infrastructure destruction. The key objectives would be establishing a correlation between traffic volume, traffic signal location with traffic congestions in urban road network during a flood event, and identification of the flood hot spots on roads and assessing potential need for interventions in urban planning. Fig. 2

Fig. 1 Flood-affected area. *(No permission required.)*

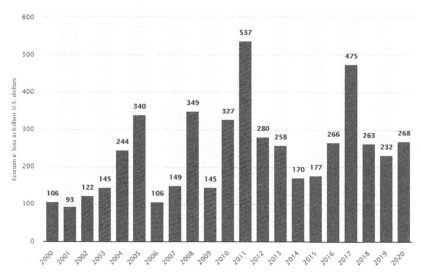

Fig. 2 Economic loss from normal disaster worldwide from the year 2000 to 2020. *(No permission required.)*

(Source: https://www.statista.com/statistics/510894/natural-disasters-globally-and-economic-losses/) shows the economic loss from normal disaster worldwide from the year 2000 to 2020.

Disaster management is very important because as of now, the global economic loss due to natural disasters was estimated to be over 268 billion USD. Natural disasters are the result of Earth's natural phenomena. Natural disasters can take many forms, including floods, earthquakes, and other natural calamities.

2. Literature survey

Due to the augmented volume of water, discarding of solid wastes and other waste into drainage, the drainage systems have been unsuccessful. Environmental fluctuations are unavoidable beings in this living world, and they are connected to every element of the ecosystem. The combination of urban flood sources demonstrates the importance of urban ecology in catastrophe risk mitigation. The information was gathered and analyzed in the context of flood risk and urban management. It demonstrates broader difficulties and provides numerous lessons for dealing with floods in our country's cities and villages. This is a review done to show various analyses and research made in the field. Comparing different kinds of methods, a solution is obtained to find and create better alternatives. The older and latest versions are compared to see if any more alterations can be made. In the existing system, only the flood management details are stressed more, from which solutions can be obtained based on them. It does not give a detailed list of the areas where the floods have occurred. It is only about finding solution for the floods to be kept in control. But this does not help always; it is only useful to control floods. The disadvantages of the existing systems are as follows: (1) testing performances are poor, (2) low accuracy, and (3) time consumption is more.

The system is designed in such a manner that water flooding can be monitored easily and managed without any difficulty [1]. A warning is generated in the system when the floods are more in that particular area. The warning will help the residents in that area to protect themselves from getting affected. The authors have emphasized a mechanism that connects the nodes to the control center, where the system has the capacity to take care of the wireless network. In the system proposed by the author, the system has network for transmission of data related to water [2]. The proposed system is designed especially for flood forecasting, allowing users to even more efficiently support WSN adaption, support broader sensor paradigms, and enable proactive behavior such as warning local stakeholders of impending flooding [3]. In the context of flood risk management, suggested a crowd-sourcing-based strategy for getting meaningful volunteer input. Volunteers are used as human sensors in this strategy, providing information about the environment such as water levels and flooded areas. The authors discovered that this platform is effective in getting relevant and accurate volunteer input depending on the preliminary validation, because volunteers can readily contribute information about the water level in the riverbed using the platform categories [4]. In this chapter, the alert is reached in the form of a notification where the notification splits and spreads immediately with the technology based on what is happening and what is being affected. The authors have also mentioned the evacuation strategies to be expanded across the area affected. All these are done by the government in the developed areas [5,6]. Early warning flood detection is like a precaution. This warning detector is done with the help of wireless sensors. To cover the regions, the system employs a minimal number of nodes. The author has also spoken about sensor network, which also deals with the requirements and communication and supports the sensor type. This chapter gives solutions to river flooding.

They created a real-time flood detection and prevention system for a specific coastal area in this research [7]. The system makes use of advanced sensing technology to provide real-time water flow monitoring. Sensor network, processing and transmission modules, and database and base station server are the three key components of the developed system. This can even be done in rural regions with access to the Internet [8]. The suggested model includes three main kinds of nodes for flood prediction and prevention using a wireless sensor network. It uses few resources while providing accurate real-time forecasts, making it an important feature in any real-world algorithm. The model has a number of parameters; that is, any type and number of components can be added or removed depending on the on-site needs [9]. The developed scheme sends a message to a specific authorized person via global communication and mobile system modem regarding water-level sensitivity. The categorization method was carried out using machine learning techniques. The research is based on analyzing flood information using a machine learning method to decide whether the water level is normal or harmful [10]. The author has focused on flood reduction in the areas where there is no outlet for water to go out. So in this case, he has used the centralized and decentralized reservoirs to help store the water and make sure the floods do not occur. In urban areas, it is most difficult for the water to be let out since the place is small and results in flood. Setting up reservoirs in those places will help them overcome these problems and also precautions to stop them [11]. The author has developed a control system, which monitors a given urban drainage control. It focuses on drains in urban areas to see if the water capacity can be handled even at the time of flood and rainfall. The control system described and presented in this chapter can avoid flash floods up to a certain extent, which is compatible with the consequences of climate change and development issues [12,13].

Munawar et al. [14] and Rahmati et al. [15] proposed a classification framework for flood risk assessment [16]. There was a shortage of optimization algorithms for flood risk assessment that included machine learning and image processing. Furthermore, the use of machine learning solutions in the aftermath of a disaster was shown to be limited [17]. The framework aided in the development of new ideas and views in the planning process and hence has the potential to affect the practice of developing flood risk reduction strategies. Access to regional, customized data on hydrological consequences was also proven to be helpful during the planning phase and when communicating with users. The technique presented here was utilized to create the second version of flood risk assessment plans as well as the third version of river basin management plans. The similar technique will be used in future planning cycles [18]. The findings of this research concentrated on two primary themes: (1) ways to society flood risk assessment for older persons and (2) characteristics that contribute to stormwater management efficiency for older adults [19,20] Conventional drainage system mitigation strategies and low-impact development (LID) practices can both effectively mitigate flooding, but a comprehensive flood management scheme emphasizing the mixture of the two has been shown to be the most successful in reducing flooding under climate change scenarios [21,22]. Flood incidents in Singapore in 2010 and 2011, particularly in the Orchard Road region, one of the city's traditionally most important retail and touristy sectors, were cited as compounding concentrating events that offer the chance for global flood management policy shifts.

These findings underscore the diversity of public attitudes about natural flood management (NFM) and the various value orientations that underpin them [23]. Sea and watershed managers who want to support NFM solutions should address public concerns regarding the usefulness of the technology in reducing flood risk and think about how to convey solutions in ways that appeal to a wide range of public values [24–26] gives an assessment of the most recent advances in flood management using image processing, artificial intelligence, and integrated techniques, with a focus on postdisaster recovery. It responds to recent artificial intelligence-based flood management strategies in a postdisaster situation, as well as current gaps in the selected technologies for postdisaster, and proposes a unique framework to optimize flood control using a holistic approach [27]. Highlights India's regional flood concerns and analyzes the steps done by the country's primary flood management agencies, with a focus on contemporary flood management strategies. The lengthy efficacy of these techniques is explored, and specific shortcomings are noted [28]. The framework is the product of a partnership between North Sea scholars and practitioners. The findings identify common difficulties and prioritize policy actions to improve "Make space for innovation" by attempting to control hazard rather than prevent it [29]. This proposed model is superior to the present flood control strategies. Decision-making and information processing have been mechanized to reduce mortality and enable stakeholders to execute relief activities more efficiently than with traditional flood disaster management approaches [30]. help us better understand the logic behind the present risk of flooding management policies and practices in different regions of the world. We investigated whether these factors explain the primary disparities reported by looking at cases related to Deltares initiatives in other countries. Suggested a checklist of criteria to examine when planning flood hazard risk management methods customized to local socioeconomic and cultural settings based on our early investigation [31]. Despite the introduction of catastrophe risk reduction measures, the analysis demonstrates that local institutions continue to use proactive methods and manage flood hazard in an ad hoc manner.

New deep learning models utilized in this study were successful in capturing the variability of spatial structure of flood likelihood in the Golestan region, and the resulting possibility maps can be used to design mitigation plans in the event of future floods [32]. The main policy consequence of our study is that flood detection systems should be designed, implemented, and verified in around 40% of the geographical area defined by high and very high flooding susceptibility [33]. A flood disaster data warehouse was created in this work using both semistructured and unstructured urban flood data. According to this, a regression model was built using a deep learning method called gradient boosting decision tree to estimate the level of urban flooded regions. Flood condition maps were created using GIS based on rainwater data from various rainfall return dates [34]. This is accomplished using a neural network design that is commonly used for picture segmentation. A detailed examination of which temporal input data should be supplied to the deep learning model and which hyperparametrization optimizes prediction accuracy and a comprehensive review of prediction accuracy for locations and rain events not regarded in training are some of the key novelties.

3. Flood detection system design

First, we have included flow diagram of flood detection system using deep learning model. In this system, aerial images are used as input information for the corresponding model, that is, deep learning model. This input information is collected through unmanned aerial vehicle (UAV) from target region. The input images can be retrieved from social media. Basically, these images are categorized into two types, which are predisaster images and postdisaster images. The variation of pre and postdisaster images is exploited as input training patches here. Both the sets of images are indistinguishable with respect to time series, for example, images captured in two different years. The system undergoes two different cases that are training and testing once the source images have been acquired from target location.

The purpose is to look for flood-related alterations in postdisaster photographs and identify them. By adding both pre and postdisaster pictures, the training stage is improved, for example, error-free. The output reveals the frequency of catastrophes by identifying disaster-related changes. The training patch has six channels because both the pre and postdisaster photographs are tri-RGB imagery. As a result, performance precision was improved with higher accuracy; refer to Fig. 3.

In this flood relief management system, the city drainage details such as length, breadth, and height are collected for some respective areas. And then, the area-wise population (number of buildings) and drain details are studied. Depending on the number of buildings, it calculates the average wastewater coming out from each and every building. The datasets for rainfall are also being collected, and depending on that, the amount of rainfall in a particular area is calculated. Combining the amount of rainfall and the wastewater, the drainage capacity of the particular area is calculated. Design consideration is very important during the development of any system. So here, we have included a high-level design for the overall drainage system as shown in Fig. 4.

3.1 Important steps involved in flood management
3.1.1 Identify flood periods
This process separates time periods where real alarm floods occur and create alarm flood sequences for those times. Depending on the water volume, the alarm logic was divided into 10-minute periods. Alarm instances above a given threshold are highlighted in periods. The threshold (t) is set given the definition of the drainage outflow for a cubic meter area.

3.1.2 Cluster flood sequences
Here, the different flood periods' area is clustered together based on a set of areas. Now, this information is clustered together to predict the formation of flood in an area. Here, the threshold value is normalized based on the cubic meter excretion of the outlet pipe from the drainage, and then, the average value is calculated and made as a clustered threshold for an area.

3.1.3 Rainfall feature extraction
Based on the amount of waterfall from the previous data for an area, we can predict the formation of flood in that area. This information shall help us to keep a minimum threshold, which can be set for rainfall and shall cause the formation of flood in that area. This calculation can be achieved by plotting a

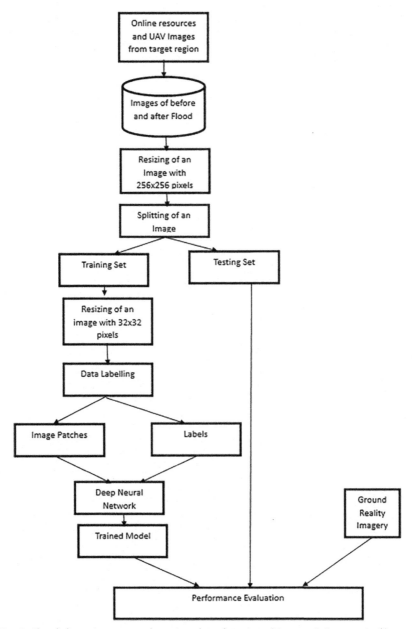

Fig. 3 Flood detection system by using deep learning. *(No permission required.)*

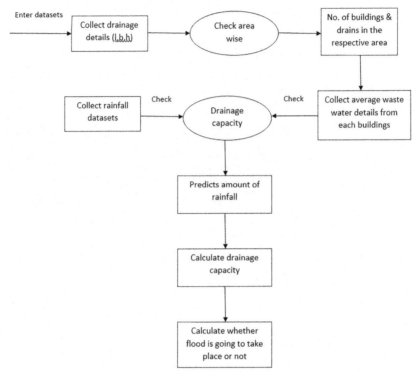

Fig. 4 High-level system design. *(No permission required.)*

graph against the amount of rainfall and the water flow in the drainage outlet. Now as we plot a graph, we can get the threshold for the next prediction for which the flood can occur for the minimum amount of rain.

3.1.4 Flood alert

Based on the prediction of flood in a certain area, we can create a flood alert for the formation of flood if the rain increases or not. This can help in preventing the flood and to take preventive measures to stop it. An alert mechanism shall be provided, which shall inform us whether the flood will occur or not. Further, refer to Fig. 5.

4. Conclusion and future work

Flooding is expected to be affected by climate change due to changes in precipitation, warmth, volume of water, and river dynamics. Climate change will worsen the current flood impacts on infrastructures and social

Fig. 5 Sample flood alert. *(No permission required.)*

organizations, such as roads, stormwater supply and wastewater systems and drainage, river flood mitigation measures, and personal and public assets, such as homes, companies, and schools. Climate change may alter flood risk assessment priorities, and in some areas, it may potentially raise the risk of flooding to dangerous levels. As a result, it is critical that flood vulnerability assessments take into account the consequences of climate change on flood hazard. Flooding is the overflow of water through any water body, such as ponds, rivers, lakes, and streams, and it can be both damaging and beneficial. Loss of property, animal deaths, and human deaths are all negative consequences. With deep learning model, it is easy to find out the flooded areas around a locality. We can find out the drains that are overflowing with valuable data, which are collected based on this model. Based on the result obtained from this model, we can take the necessary precautions to stop floods from occurring in an area. Likewise, the affected areas and overflowing drainages can be identified. Due to the presence of numerous hydro–climatological elements, the flood management procedure in India is extremely complicated. Every year, floods cause significant harm to life and property, so it is time for the federal and state governments to develop a long-term plan that goes beyond flood control measures like embankments and dredging. There is also a pressing need for a coordinated basin management strategy that connects all river basins to drains. In the future, alternative treatments such as a mixture of minimizing the time where possible, attempting to control hazards through structural measures, consequences of adverse events via insurance, accepting risk, emergency preparedness planning, detection systems, and communicating risk to affected parties could be implemented. A mobile application can be developed in the future to alert people with a message. This would be easy and helpful for all people residing in the area. Natural flood management or even working with natural process will help to reduce the flood and risk in urban areas where implementing measures that help to protect, restore everything. Natural flood management will help reduce maximum water floods and delay floods.

References

[1] J. Sunkpho, C. Ootamakorn, Real-time flood monitoring and warning system, Songklanakarin J. Sci. Technol. 33 (2) (2011) 227–235. http://www.rdoapp.psu.ac.th/html/sjst/journal/33-2/0125-3395-33-2-227-235.pdf.

[2] D. Hughes, P. Greenwood, G. Blair, G. Coulson, P. Smith, K. Beven, An Intelligent and Adaptable Grid-based Flood Monitoring and Warning System, CiteSeer, 2006.

[3] L.C. Degrossi, J.P. De Albuquerque, M.C. Fava, E.M. Mendiondo, Flood citizen observatory: a crowdsourcing-based approach for flood risk management in Brazil, in: Proceedings of the International Conference on Software Engineering and Knowledge Engineering, SEKE, vol. 2014 (January), Knowledge Systems Institute Graduate School, 2014, pp. 570–575. http://www.ksi.edu/seke/seke13.html.

[4] M. Usha, P. Kavitha, Anomaly based intrusion detection for 802.11 networks with optimal features using SVM classifier, Wirel. Netw. 23 (8) (2017) 2431–2446, https://doi.org/10.1007/s11276-016-1300-5.

[5] E.A. Basha, S. Ravela, D. Rus, Model-based monitoring for early warning flood detection, in: SenSys'08—Proceedings of the 6th ACM Conference on Embedded Networked Sensor Systems, 2008, pp. 295–308, https://doi.org/10.1145/1460412.1460442.

[6] C. Rajeswari, B. Sathiyabhama, S. Devendiran, K. Manivannan, Bearing fault diagnosis using wavelet packet transform, hybrid PSO and support vector machine, in: Procedia Engineering, vol. 97, Elsevier Ltd., 2014, pp. 1772–1783, https://doi.org/10.1016/j.proeng.2014.12.329.

[7] A.A. Pasi, U. Bhave, Flood detection system using wireless sensor network, Int. J. Adv. Res. Comput. Sci. Softw. Eng. 5 (2015) 386–389.

[8] G. Megha, S. Manojkumar, P. Babasaheb, D. Akshay, Early flood detection system using android application, Int. J. Eng. Res. Technol. 8 (2019) 681–686.

[9] M. Khalaf, P. Fergus, I.O. Idowu, Advance flood detection and notification system based on sensor technology and machine learning algorithm, in: IEEE, International Conference on Systems, Signals and Image Processing, 2015.

[10] E. Lee, Y. Lee, J. Joo, D. Jung, J. Kim, Flood reduction in urban drainage systems: cooperative operation of centralized and decentralized reservoirs, Water 8 (10) (2016) 469, https://doi.org/10.3390/w8100469.

[11] J. Leitao, A. Cardoso, J.A. Marques, N. Simoes, Flood Management in Urban Drainage: Contributions for the Control of Water Drainage Systems using Underground Barriers, IEEE, 2018, pp. 21–28.

[12] N.S. Hsu, C.L. Huang, C.C. Wei, Intelligent real-time operation of a pumping station for an urban drainage system, J. Hydrol. 489 (2013) 85–97, https://doi.org/10.1016/j.jhydrol.2013.02.047.

[13] S. Mugume, D.E. Gomez, D. Butler, Quantifying the resilience of urban drainage systems using a hydraulic performance assessment approach, in: Proceedings of the 13th International Conference on Urban Drainage, IAHR, 2014, pp. 7–11.

[14] H.S. Munawar, A.W.A. Hammad, S.T. Waller, A review on flood management technologies related to image processing and machine learning, Autom. Construct. 132 (2021) 103916, https://doi.org/10.1016/j.autcon.2021.103916.

[15] O. Rahmati, H.R. Pourghasemi, H. Zeinivand, Flood susceptibility mapping using frequency ratio and weights-of-evidence models in the Golastan Province, Iran, Geocarto Int. 31 (1) (2016) 42–70, https://doi.org/10.1080/10106049.2015.1041559.

[16] P.-T. Ngo, N.-D. Hoang, B. Pradhan, Q. Nguyen, X. Tran, Q. Nguyen, V. Nguyen, P. Samui, D. Tien Bui, A novel hybrid swarm optimized multilayer neural network for spatial prediction of flash floods in tropical areas using sentinel-1 SAR imagery and geospatial data, Sensors 18 (11) (2018) 3704, https://doi.org/10.3390/s18113704.

[17] A. Parjanne, A.-M. Rytkonen, N. Veijalainen, Framework for climate proofing of flood risk management strategies in Finland, Water Secur. 14 (2021).

[18] P. Yodsuban, K. Nuntaboot (2021). Community-based flood disaster management for older adults in southern of Thailand: A qualitative study, Int. J. Nurs. Sci. 8, 4079-417.

[19] M. Usha, P. Kavitha, Anomaly based intrusion detection for 802.11 networks with optimal features using SVM classifier, Wirel. Netw. 23 (8) (2017) 2431–2446, https://doi.org/10.1007/s11276-016-1300-5.

[20] X. Sun, R. Li, X. Shan, X.H. Xu, Assessment of climate change impacts and urban flood management schemes in central Shanghai, Int. J. Disaster Risk Reduct. 65 (2021).

[21] K.S. Loh, M.D. Pante, Controlling nature, disciplining human nature: floods in Singapore and Metro Manila, 1945–1980s, Nat. Cult. 10 (1) (2015) 36–56, https://doi.org/10.3167/nc.2015.100103.

[22] C. Tortajada, R. Koh, I. Bindal, Compounding focusing events as windows of opportunity for flood management policy transitions in Singapore, J. Hydrol. (2021) 599.

[23] M. D'Souza, M.F. Johnson, C.D. Ives, Values influence public perceptions of flood management schemes, J. Environ. Manag. 291 (2021), 112636, https://doi.org/10.1016/j.jenvman.2021.112636.

[24] H.S. Munawar, A.W.A. Hammad, S.T. Waller, M.J. Thaheem, A. Shrestha, An integrated approach for post-disaster flood management via the use of cutting-edge technologies and UAVs: a review, Sustainability 13 (14) (2021) 7925, https://doi.org/10.3390/su13147925.

[25] Scottish Government, Scottish Government Urban Rural Classification, 2016. https://www.gov.scot/publications/scottish-government-urban-rural-classification-2016/.

[26] J. Vávra, M. Lapka, E. Cudlínová, Z. Dvořáková-Líšková, Local perception of floods in the Czech Republic and recent changes in state flood management strategies, J. Flood Risk Manag. 10 (2) (2017) 238–252, https://doi.org/10.1111/jfr3.12156.

[27] M.P. Mohanty, S. Mudgil, S. Karmakar, Flood management in India: a focussed review on the current status and future challenges, Int. J. Disaster Risk Reduct. 49 (2020) 101660, https://doi.org/10.1016/j.ijdrr.2020.101660.

[28] P. Sayers, B. Gersonius, F. den Heijer, W.J. Klerk, P. Fröhle, P. Jordan, U.R. Ciocan, J. Rijke, B. Vonk, R. Ashley, Towards adaptive asset management in flood risk management: a policy framework, Water Secur. 12 (2021) 100085, https://doi.org/10.1016/j.wasec.2021.100085.

[29] H.R. Goyal, K.K. Ghanshala, S. Sharma, Post flood management system based on smart IoT devices using AI approach, Mater. Today: Proc. 46 (2021) 10411–10417, https://doi.org/10.1016/j.matpr.2020.12.947.

[30] F. Klijn, M. Marchand, K. Meijer, H. van der Most, D. Stuparu, Tailored flood risk management: accounting for socio-economic and cultural differences when designing strategies, Water Secur. 12 (2021), 100084, https://doi.org/10.1016/j.wasec.2021.100084.

[31] I.A. Rana, M. Asim, A.B. Aslam, A. Jamshed, Disaster management cycle and its application for flood risk reduction in urban areas of Pakistan, Urban Clim. 38 (2021), 100893, https://doi.org/10.1016/j.uclim.2021.100893.

[32] M. Panahi, A. Jaafari, A. Shirzadi, H. Shahabi, O. Rahmati, E. Omidvar, S. Lee, D.T. Bui, Deep learning neural networks for spatially explicit prediction of flash flood probability, Geosci. Front. 12 (3) (2021) 101076, https://doi.org/10.1016/j.gsf.2020.09.007.

[33] Z. Wu, Y. Zhou, H. Wang, Z. Jiang, Depth prediction of urban flood under different rainfall return periods based on deep learning and data warehouse, Sci. Total Environ. 716 (2020) 137077, https://doi.org/10.1016/j.scitotenv.2020.137077.

[34] R. Löwe, J. Böhm, D.G. Jensen, J. Leandro, S.H. Rasmussen, U-FLOOD—topographic deep learning for predicting urban pluvial flood water depth, J. Hydrol. 603 (2021) 126898, https://doi.org/10.1016/j.jhydrol.2021.126898.

Further reading

A.K. Biswas, Droughts or Floods: what is important for Singapore? 2012. https://lkyspp.nus.edu.sg/gia/article/droughts-or-floods-what-isimportant-for-singapore.

CHAPTER FOUR

Powering data-driven decision-making for the development of urban economies in India

Radha Karmarkar[a], Aman Singh Rajput[b,c], and Vivin R. Nair[a]
[a]India Smart Cities Fellowship (ISCF), National Institute of Urban Affairs under the Ministry of Housing and Urban Affairs, New Delhi, India
[b]Urban Infrastructure and Tourism, IPE Global Limited, New Delhi, India
[c]India Smart Cities Mission, New Delhi, India

1. Overview

The 74th Constitutional Amendment Act of Government of India in 1992 gave the provision for devolution of the function "socio-economic planning" from the state to the urban local bodies. This function is undertaken by the city managers through the preparation of the Master/Development Plan drawn up by a statutory directive through the state Town and Country Planning Acts [1,2]. The methodology of master plan preparation is static and has a unilateral approach, which impedes the city from reacting positively and proactively to address the restructuring needs and meeting national competition [3]. There are financial plans and commissions at the central and state level that focus on economic planning, for example, NITI Aayog prepared the Five-Year Plans or State Finance Commissions prepare the State Finance Commission Reports showing the status of municipal finance; yet there are no economic plans at the urban local body level, which have some of the largest economic agglomerations in the country. Furthermore, there is no single source of information and analysis for city authorities and investors to know the viable sectors for economic growth and the forward-backward linkages between them. Apart from limited functions and knowledge resources, cities also have limited data or data repositories to consolidate and clean economic data. Hence, economic decisions are taken via a top-down approach to benefit the larger central and state goals. Urban local bodies should be enabled to take a data-driven approach for policy making and planning, which is informed by research and analysis that is suited to the problems identified. In this backdrop, this chapter details out the methodology prepared for the development of Local Economic Intelligence Platform (LEIP)—a project conceived by the Ministry of Housing and Urban Affairs under the India Smart Cities Fellowship Program 2020.

2. Literature review

2.1 Best practices and the complexity discourse

One of the most significant works in analysis of the entire economy of a region is the "Atlas of Economic Complexity" written by Ricardo Hausmann. In this book, Hausmann shows how countries with greater economic complexity tend to have faster economic growth as compared to the economies that are "too rich for their current level of economic complexity." In simple terms, the book describes economic complexity as the amount of knowledge in the productive sector of the economy. Hausmann says that economic complexity is the driver of economic prosperity. The book forms an Economic Complexity Index, which countries can use to understand the complexity of products they successfully export and the Complexity Outlook Index, which countries can improve by moving closer to parts of the product space that have more complex products. The project takes inspiration from this book to understand the economic complexity of a region, and in this case, a city. Since this project had aimed to understand the various economic actors, the capital available, products manufactured, and labor in the city, we also found frameworks for understanding the economic complexity [4].

One study by Hausmann on complexity was on the country of Panama. The paper analyses the exports of the country to understand strategic export diversification opportunities for Panama. The paper uses the empirical approach to predict the growth rate of different industries in the economy. Industries of the economy and their reliance on the on-ground productive capacities and the demand for skills are analyzed to aid policy recommendations [5].

The "Competitiveness of Cities" report published by the World Economic Forum in 2014 creates a "Four-Part Taxonomy of City Competitiveness." This includes institutions, policies, and regulation of the business environment, hard connectivity, and soft connectivity. A "big basket" of 33 cities and a "small basket" of 26 cities located on all continents of the world were analyzed according to this taxonomy. The report states that based on the analysis on what to reform and how to reform in the institutional setup of the cities, policies and regulations on businesses, core physical infrastructure, and soft connectivity. The report threw light on the importance and the relation of technology, flexibility to adapt and reform at the municipal level and "successful" cities [6].

In order to enable cities to form economic development plans, the Asian Development Bank (ABD) created a toolkit that provides a framework for tools to conduct rapid economic assessment in Asian cities. The toolkit, titled "Toolkit for Rapid Economic Assessment, Planning and Development of Cities in Asia," guides the user to prepare economic profiles, evaluate economic development pathways, and prepare strategies and action plans for cities in Asia. It gives various qualitative and quantitative assessment tools for investors, government officials, community leaders, and other civil society agencies for informed decision-making. The authors of the toolkit state that the reason behind preparing the toolkit is that Asian cities have limited resources, capacity, and other limitations for planning for the local economy. Additionally, other guides and tools for such economic planning are designed for cities in developed countries and cannot directly be used in Asian countries. Other than providing guidance on aggregation of data, conducting rapid assessments, and preparing development plans, this toolkit also provides methods of preparing projections of the city through expected population, economic growth, and resources demanded. It helps the city to understand its future risks and opportunities [7]. The LEIP project also relies on this toolkit and framework as a base to assessing economic complexity.

While the Atlas of Economic Complexity and the various indices mentioned in it centered largely on the productivity and products, the team's research also included the factors that affect the productivity in a region, like education, technology, labor, and public policy.

2.2 Factors in the local economy

An article named "Regions and Universities Together Can Foster a Creative Economy" by Richard Florida, based on the book "Creative Economy" by the same author, assesses the role of university, i.e., education in the 3 Ts of economic growth—technology, talent, and tolerance. The findings of the paper suggest that education not only plays a part in fueling commercial technology, but also attracting and mobilizing talent. A "university" is a creative hub for a regional economy, and it stimulates the spillover of technology, talent, and tolerance into a community. Universities also attract the top talent from the resource pool. By attracting students from various ethnic and racial backgrounds, income levels, and sexual orientations, they give rise to tolerance and generate new ideas, innovations, and entrepreneurial enterprises, which positively impact economic growth [8].

A working paper on "Infrastructure and City Competitiveness in India" examines how the infrastructure supply in a city affects the competitiveness of the city. The paper collects data from Indian cities on the infrastructure provision and the degree of private investment in the cities. The analysis of data shows that the closer the city is to any international port or highways that connect to large domestic markets, the larger effect it has on private investment in the city. The paper also shows that local infrastructure services do enhance competitiveness but have a comparatively lesser effect than international ports or national/state highways. It talks about municipal finances as well in the wake of infrastructure services. It shows that when cities increase their municipal revenue through local taxes and user fees, infrastructure supply increases as opposed to increase in revenue through intergovernmental transfers [9].

2.3 Economic clusters and agglomeration

In a paper that studies the economic growth of metropolitan cities in India since 1991, taking the cases of Kolkata and Bangalore showed the differences in governance practices, investment patterns, and growth with the background of local exigencies. This paper titled "Metropolitan City Growth and Management in Post-Liberalized India" studies the policy changes after liberalization, investment patterns in major metropolitan areas between the years 1995 and 2010, and the history of governance in Kolkata and Bangalore. The paper shows that reform-related progress has been higher in economically dynamic regions [10].

Agglomeration present in cities propels it toward growth by attracting businesses. With new urban missions, like the Smart Cities Mission, cities are attracting investments through public-private partnerships (PPPs). Some studies have indicated that India should invest more than $150 billion in the next 5 years [11]. A paper titled "PPP Experiences in Indian Cities: Barriers, Enablers and Way Forward" studies the kind of barriers PPP projects face in India. These barriers include lack of political will, absence of enabling institutional environment for PPPs, lack of project preparation capacity, and poorly designed and unstructured PPP projects. This shows that although PPPs are necessary, the importance for the same needs to be stressed at the political and administrative level in India [11].

A JP Morgan and Chase and Co (JPMC) report on Building Strong Clusters says that cities need clusters to catalyze economic growth. The report says that many cities have struggled to accelerate the growth of targeted clusters that have the potential to drive future economic growth. The report takes examples of cluster growth in cities like San Diego and Chicago and gives recommendations for designing high-impact cluster growth strategies, and it stresses on the importance of developing strong PPPs [12].

3. AI/ML in local economy: Problem statement

As mentioned in the first section, there is a lack of planning at the city level in India, while there are economic and financial plans made at the National and State level. Districts are also mandated to make District Development Plans. Indian cities are centers of economic agglomeration. Despite this, there are no sources with aggregated economic data at the city level.

Agglomeration in Indian cities offers unique opportunities for the international as well as domestic market. This involves supplementing national and state level policies with bottom-up agglomeration level intervention where economic activities are spatially mapped. There is a need to create economic powerhouses based in cities for manufacturing of basic goods and services based on the localities' comparative advantage, leveraging agglomeration effects and easy local logistics avoiding iceberg transportation costs. The platform is intended for the use of government officials concerned with urban development to analyze, infer, and understand the economic intricacies of the city on a visual and spatial scale and provide them with a decision support system to facilitate positive interventions in economic development.

Public-private partnerships (PPPs) for urban infrastructure face many hurdles like inadequate rate of return, distrust between private and public sectors, lack of political will toward project implementation, lack of enabling institutional environment, and also, lack of public sector capacity to select and procure PPP projects. There is no aggregated platform for investors to know which city will give them better return on investment, or for the ULB to make economic plans. Neither is there any platform for entrepreneurs to find out how economically conducive environment the city has. Such an aggregated platform can help investors approach a city, as this can also be beneficial for a city to increase their own investment potential through tax and charges such as property tax, electricity, and water charges.

To increase the growth of cities, they should be aware of the gaps and different opportunities in their local ecosystems to attract investment. As there is no tool for cities to trace the annual income, economic linkages, and identifying economic clusters within the city, cities are not able to prepare an economic plan as there is lack of data on all of the above. Additionally, Urban local bodies need to generate revenue to provide basic services to its citizens. As the city grows economically, the citizens should ideally benefit directly from it. While mapping the economic opportunities in the city, another layer can be added to map the priority sectors with revenue potential in the city given the land resources it has. This project aims to assess the comparative advantage of the city based on the sectors and industry it houses and its capacity to garner more investment. This can reduce the information asymmetries between investors and city/state governments.

A part of the solution is not only aggregation and analysis of economic data in the city, but also building a platform that uses machine learning to use time series data, establish correlations between different variables, and generate a suggestive roadmap for cities to aid decision-making. Such a suggestive roadmap and predictive analysis based on existing data can help the city make economic development plans.

4. LEIP: Introduction and methodology

The product developed at the India Smart Cities Fellowship Program was ideated to incorporate scalability and replicability to the product.

The platform aggregates and analyses data at the local level to provide economic intelligence to the administrators, researchers, and investors. LEIP uses an in-house developed Local Economic Development framework to understand where the local economy stands and places the city on a four-pillar axis. This is based on an index score calculated from 100 indicators of the framework. The platform does a Rapid Economic Assessment of the economy based on the four pillars. It identifies the comparative and competitive advantage of the local economy through location quotient and shift-share analysis. Based on this analysis, the platform will provide economic intelligence to the urban authorities, that is, identify the gaps in the ecosystem of the economy and give direction for intervention. The platform will also do cluster mapping and cluster analysis by identifying the forward and backward linkages of different sectors of the economy and thereby provide insights about sectors with growth potential.

4.1 Rapid economic assessment

Rapid Economic Assessment is an evaluation of the city across the four pillars of the economy: human capital, infrastructure, investment, and local policy. Each of these pillars has a set of indicators and variables associated with them. With the help of data given for each of these indicators, an index is calculated for each of these pillars. The methodology for the same is given in Box 1.

BOX 1

Each Pillar is subdivided into different categories or variables under which there are indicators characterizing the variables. The methodology in building an index for human capital, investment, and infrastructure is given below:

Primary problem

The variables under each pillar are of different measurement or scale, which makes it complex for comparison purposes or summing them to build an index. The primary problem to solve is to rescale the indicators under each category variable of a pillar to a similar scale in order to have the addition of variables possible.

Standardization or normalization technique

To overcome this difficulty of different types of measurements on indicators of each category variable under a pillar, the following transformation is given to the indicator value

$$Z = (X - \mu)/\sigma$$

where X is the value of the indicator, μ is the mean value of the indicator, and σ is the standard deviation of the indicator. The step is required to make the indicators comparable with each other. This transformation is called normalization or standardization procedure. Through this procedure, we generate a Z score, which enables us to compare the indicators as the indicators are now in a similar scale. The value of Z score usually lies between -3 and 3. For a large number of values, the Z variable follows standard normal distribution (SND) with zero mean and one as standard deviation, which is an important theoretical distribution in statistical theory. The theory supports that

About 68.28% of observation on a random variable following SND falls under -1 to 1.

About 95.46% of observation on a random variable following SND falls under -2 to 2.

About 99.73% of observation on a random variable following SND falls under -3 to 3.

Thus, the probability that a value on the random variable X lies outside -3 to 3 is as low as 0.0027. It is critical to normalize the data before making any data aggregation as indicators have different units. The normalization procedure is carried out to transform all the data into dimensionless numbers. This is done using Z scores that can be placed in a normal distribution.

Aggregation on category variable

After computing the Z score on each indicator using standardization technique, we compute a score for the category variable, namely, category score, using the formulae of weighted average. That is, the statistical tool used for computing category score is weighted average and is given by:

$$\text{Category Score} = \sum (i = 1)^n W_i \times Z_i \text{ Score}$$

Continued

BOX 1—cont'd

where W_i represents the weightage of ith indicator and Z_i Score is the Z score of the ith indicator. Here, "n" represents the number of indicators under a category variable. Thus, a category score for each category variable with a certain number of indicators is generated.

Rescaling the category score from 0 to 100

Now, we rescale the generated category score into a 0 to 100 scale with an intention to keep the score similar to an index (index usually lies in a range 0 to 100). The rescaling of the category score of each category variable into 0 to 100 range, we use the following formula:

$$\text{Rescale Score} = (\text{Category Score } (i) - \text{MinimumScore})/(\text{Maximum Score} - \text{Minimum Score})$$

Here, the Category Score (i) represents the ith category score of the pillar. The minimum and maximum scores are the lowest and highest scores among all the category score of a pillar, respectively. Thus, a rescale score varies from 0 to 100 for each category variable is obtained under a pillar.

This component provides an overview of the economy and shows the gaps within the economic ecosystem. A higher index score on either of the pillars will show that the city has adequate interventions for those pillars, whereas

The Rapid Economic Assessment gives an overview of the city economy and also shows the areas of intervention according to the pillar index. For example, if the index score for one of the pillars, human capital is low, then the city administrators can conclude that they need to make policies and investments in the human capital of the city. They can increase resources in improving the skills and education or any other human capital indicator to increase the index score for that pillar. The indicators used in REA are reflective of the complexity in the economy and the more the number of indicators, the more complexity will be captured by the tool leading to better accuracy of results.

4.2 Economic competency

A thorough understanding of the local economy helps highlight the viable sectors most suitable for growth in the city. The platform uses the measures of location quotient (LQ) and shift-share analysis (SSA) to provide information on the comparative and competitive advantage of the city's economy. Directory of Establishments data under the Economic Census conducted by the Ministry of Statistics and Programme Implementation (MoSPI) formed the base of this analysis.

Location Quotient: The study of a city's competitiveness starts with the simple location quotient, which does not require extensive data and processing. It is a simple tool for comparing a region's share of a certain activity with its share of some basic aggregate. In the three-tier Indian political system, a city's economy is more closely related to that of the state in which it is located; hence, the state is chosen as the reference area for the city in the tool. The advantage of the LQ method is its simplicity and that it is based on readily available data. Policy makers may be interested in how employment in industries A and B is concentrated across regions of the nation. Regional policy makers should know which industries are concentrated in their own regions.

If LQ > 1 means that the industry employs a greater share of local workforce than it does nationally (or in the state), which implies that the industry is producing more than is consumed locally. If LQ < 1, then the goods we are examining in the region would be an import industry, since it means that local residents and businesses are purchasing these particular goods and services from outside the local area. When examining changes in LQs for the same region over time, keeping in mind its limitations which are discussed later, large declining LQs over time would indicate that the industry is important to the local economy and losing it would create problems. Small and growing LQs over time would indicate that the industry will promise future growth for the local economy and that it should be supported. Small declining LQs indicate that they are not important to the local economy; large increasing LQs are desirable since they are the base of the local/regional economy. Analysis of the region/city's LQs for all industries would therefore send signals as to what the area's competitive advantage is and how it is changing over time, then the good we are examining in the region would be an import industry, since it means that local residents and businesses are purchasing these particular goods and services from outside the local area. Small and growing LQs over time would indicate that the industry will promise future growth for the local economy and that it should be supported. Small declining LQs indicate that they are not important to the local economy; large increasing LQs are desirable since they are the base of the local/regional economy. Analysis of the region/city's LQs for all industries would therefore send signals as to what the area's competitive advantage is and how it is changing over time.

Shift-share analysis (SSA): SSA examines economic growth or decline, measured by employment (or other indicators such as value added or output) in a city by disaggregating the change into three components: (1) National share, (2) Industrial mix, and (3) Regional share. The measurement unit could be employment, income, output, value added, or other factors. In this toolkit, employment is used due to the availability of this data at the city level from the Census of India and MoSPI.

National share (NS)—This part of the shift-share technique measures the employment change that occurred in the city, if its employment grew at the same rate as that of the state (which is considered as the reference area in this paper). In other words, this part shows the city's employment change during the previous period that was due to national economic growth, which invariably influences its constituent economies, here the city. Since we are examining cities here, the state economy is a significant part of the regional economic environment the city is faced with. If the state's (reference area) economy had been experiencing high overall macroeconomic growth, the city's employment would grow as well.

Industry mix (IM)—This component of the shift-share technique reflects the share of local economic (employment) change that is due to the city's industry composition. If the city contains some industries that are fast growing nationally, then it could experience faster employment growth in those industries and vice versa. Positive net values for IM indicate that the industry composition of the city contains faster growing industries. Negative results indicate the opposite.

Regional share (RS)—This component measures the employment change in a particular sector of the city as being the difference between the sector's growth rate in the city and the sector's growth rate in the state. This component is the one that indicates the local area's competitiveness in producing a particular good or service. This could be a natural resource, skill availability, size of a market, local consumption, savings, institutions, governance, or local leadership.

What is termed as the "total shift" in shift-share analysis is thus the sum of the three components (NS, IM, and RS) and is equal to the employment change in the city over the said period for a specific sector.

Concurrence table for the economic census data comparison

Division	Activity	Division
2008		2004
1	Crop and animal production, hunting and related service activities	1
2	Forestry and logging	2

Concurrence table for the economic census data comparison—cont'd

Division	Activity	Division
3	Fishing and aquaculture	5
5	Mining of coal and lignite	10
6	Extraction of crude petroleum and natural gas	11 (1)
9	Mining support service activities	11 (2)
	OMQ	12
7	Mining of metal ores	13
8	Other mining and quarrying	14
10	Manufacture of food products	15 (1)
11	Manufacture of beverages	15 (2)
12	Manufacture of tobacco products	16
13	Manufacture of textiles	17
14	Manufacture of wearing apparel	18
15	Manufacture of leather and related products	19
16	Manufacture of wood and of products of wood and cork, except furniture; manufacture of articles of straw and plaiting materials	20
17	Manufacture of paper and paper materials	21
18	Printing and reproduction of recorded media	22 (1)
58	Publishing activities	22 (2)
19	Manufacture of coke and refined petroleum products	23
20	Manufacture of chemicals and chemical products	24
22	Manufacture of rubber and plastics products	25
23	Manufacture of other non-metallic mineral products	26
24	Manufacture of basic metals	27
25	Manufacture of fabricated metal products, except machinery and equipment	28
28	Manufacture of machinery and equipment n.e.c	29
27	Manufacture of electrical equipment	31
59	Motion picture, Video and television programme production, sound recording, and music publishing activities	32
21	Manufacture of pharmaceuticals, medicinal chemical, and botanical products	33 (1)
26	Manufacture of computer, electronic and optical products	33 (2)
	MOC, E, and OP	30
29	Manufacture of motor vehicles, trailers, and semitrailers	34
30	Manufacture of other transport equipment	35
31	Manufacture of furniture	36 (1)
32	Other manufacturing	36 (2)
38, 39	Waste collection, treatment, and disposal activities; materials recovery; remediation activities; and other waste management services	37
35	Electricity, gas, steam, and air conditioning supply	40
36	Water collection, treatment, and disposal	41

Continued

Concurrence table for the economic census data comparison—cont'd

Division	Activity	Division
41,42	Construction of buildings, civil engineering	45 (1)
43	Specialized construction activities	45 (2)
45	Wholesale and retail trade and repair of motor vehicles and motorcycles	50
46	Wholesale trade, except of motor vehicles and motorcycles	51
47	Retail trade, except of motor vehicles and motorcycles	52
55	Accommodation	55 (1)
56	Food and beverage service activities	55 (2)
49	Land transport and transport via pipelines	60
50	Water transport	61
51	Air transport	62
52	Warehousing and support activities for transportation	63 (1)
79	Travel agency, tour operator, and other reservation service activities	63 (2)
53	Postal and courier activities	64 (1)
61	Telecommunications	64 (2)
64	Financial service activities, except insurance and pension funding	65
65	Insurance, reinsurance and pension funding except compulsory social security	66
66	Other financial activities	67
68	Real estate activities	70
77	Rental and leasing activities	71
62	Computer programming, consultancy, and related activities	72 (1)
63	Information service activities	72 (2)
72	Scientific research and development	73
69	Legal and accounting activities	74 (1)
70	Activities of head offices; management consultancy services.	74 (2)
71	Architecture and engineering activities; technical testing and analysis	74 (3)
73	Advertising and market research	74 (4)
74	Other professional, scientific, and technical activities	74 (5)
78	Employment activities	74 (6)
80	Security and investigation activities	74 (7)
81	services to buildings and landscape activities	74 (8)
82	Office administrative, office support, and other business support activities	74 (9)
85	Education	80
75	Veterinary activities	85 (1)
86	Human health activities	85 (2)
87	Residential care activities	85 (3)
88	Social work activities without accommodation	85 (4)
37	Sewerage	90

Concurrence table for the economic census data comparison—cont'd

Division	Activity	Division
94	Activities of membership organization	91
60	Broadcasting and programming activities	92 (1)
90	Creative, art, and entertainment activities	92 (2)
91	Libraries, archives, museum, and other cultural activities	92 (3)
92, 93	Gambling and betting activities; sports activities; and amusement and recreation activities	92 (4)
84	Public administration and defense; compulsory social security	75
96, 97	Other personal service activities; activities of households as employers of domestic personnel	95
	OPSA, AOHAEODP	96

4.3 Cluster analysis

The cluster mapping and analysis is incorporated into the tool to understand the economic geography of the city and its interconnectedness. The idea was to interpret the economic complexity and diversity of the city. The dataset used for the analysis is from the Goods and Service Tax (GST). The literature review was carried out to understand the various categories of forms used by the businesses to file the income tax for various types of sales and purchase from business to business and business to customer transactions. While discussing with the stakeholders from the industry, business to customer's dataset was too large in volume and using it to build the backend would get difficult. Therefore, the team decided to go ahead with the business to business transactions taking place in the region using the consolidated GSTR 9 and GSTR 2A data along with the e-way bills. The data at business level transactions involve personal information that cannot be shared in the public domain, and therefore, the methodology for analysis is prepared as follows:

(1) The data on the number of establishments were collected not from the exact address of the businesses but the pin code at which they are present. This would enable the system to understand which type of businesses are located or clustered at a specific location. The dataset received from State GST Bhavan was overlaid on the pin code-wise map of the district. Furthermore, if required the pin codes can be further subdivided into rural and urban counterparts.

(2) Sectoral wise outward supply turnover dataset was collected and interpreted using the bar graph to understand the export potential of the sectors of the city and its related turnover. Further, the payable tax data under the three heads of CGST (Central GST), SGST (State GST), and IGST (Integrated Goods and Services Tax) were provided by the GST Bhavan to facilitate the outward supply within the state (adding CGST and SGST) and outside the state (IGST). This helps in the interpretation of the export potential of the city within the state and at the national level.

(3) Sectoral distribution of inward supply turnover was collected using the e-way bill data from the assessable value of the goods moving inside the city at two levels inter-state and intra-state. This provides information about the goods that are consumed by the city from within the state and from the national level.

(4) Sectoral distribution of import and export turnover using the assessable value of the e-way bills is used to compare the goods movement inside and outside the region. Analyzing these simultaneously helps in the interpretation of the services sector potential for growth.

Sectoral distribution of export and import turnover was analyzed using the arc diagram to interpret the forward and backward linkages of the economy. The business-to-business transactions between two sectors represent their association. This would enable the policymakers to interpret the interconnectedness of the economy and analyze the multiplier effect of different economic sectors.

An analysis of the GST data gives a picture of the businesses present in the city's economy and their trade with other businesses. Like mentioned before, cities are hubs of agglomeration, and hence, knowledge of business to business transactions gives an overview of the forward and backward linkages in the economy. This tab also helps understand if the city is a producer of goods or a consumer of goods and its relationship with other states and cities and the dependencies involved.

4.4 Decision support system

In order to aid the process of decision-making for administrators and local authorities, the platform provides particulars on the potential economic sectors and economic resiliency of the city and provides intervention direction for the viable sectors of the urban economy. Further, it would facilitate greater growth and human development—placing the city on a self-sustaining, positive growth trajectory that matches its potential growth. The methods used for processing the said information are conditional rules upon the values of economic competency. Additionally, gross value added and foreign direct investment datasets are used for interpreting the economic dependency through condition based percentage level. Finally, the aforementioned functionalities of the platform recommend the positive intervention direction and diversification to the users.

The potential sectors of the economy are identified using the conditional ruling on the values of location quotient, shift share, and share of employment of the economic sectors. These conditions are not rigid and could be modified as per the city requirement and context. Location quotient represents the competitiveness of the sector, shift share represents the sector's economic growth share, and share of employment represents the human capital holding capacity of the sectors. Based on the permutations and combinations of the value ranges of these parameters, the platform identifies the viable sector of development.

Viable Class I	Viable Class II	Viable Class III	Viable Class IV
LQ > 1	LQ < 1 and difference between LQ 2004 and 2013 is (+)ve or LQ < 1 and share of employment >5%	LQ < 1 and either of industrial mix and regional share is (+)ve	Others
1.1: LQ 1 to 1.5			
1.2: LQ 1.5 to 2			
1.3: LQ 2 to 3			
1.4: LQ > 3			

The platform calculates the resiliency of the economy based on the GVA share and the FDI inflows of various economic sectors. The GVA share of the top 5 sectors of the economy is added and based on its value in 5 classes over a scale of 0 to 100, resiliency of the city economy is studied from very low to very high. The following five-scale classification had been identified by the authors in consultation with the field experts and stakeholders.

Cities with top 5 sectors GVA share	Resilience in the economy
0%–35%	Very high
35%–50%	High
50%–65%	Moderate
65%–80%	Low
80%–100%	Very low

Furthermore, the platform provides information on the individual economic sector performance on the four axis pillar of the rapid economic assessment as well as the intervention direction required under each.

The methodology for calculation of individual sector performance on infrastructure, investment, human capital, and policy variable involved comparison with other cities' performance and the variability of importance for similar infrastructure impacts on the development of the specific economic sectors. Thus, the first step involved the identification of the variables under each pillar, whose dataset pertaining to the specific economic sector is available. Further, the weightages for each of the variables were defined based on the variable's role in the development of the sector. Finally, statistical methods of indexing were used to place the city's particular economic sector on the four axis values of REA.

5. Envisioning AI/ML in LEIP and future of local economic planning

Going ahead, a platform like this can be integrated with artificial intelligence to make it more complex. Time series data can be included for conducting the rapid economic assessment and sector-wise analysis to analyze the gaps in the ecosystem. With the help of machine learning, this platform can utilize the data to provide a suggestive roadmap to the cities. One of the aims of this tool is to equip ULBs and other stakeholders involved in city level planning with economic data and analysis so that they can make informed decisions. This information should also be converted to economic development plans according to which policy actions can be taken for the benefit of the city. The current platform provides 128 datasets from approximately 30 departments in the city of Thiruvananthapuram. These datasets have been standardized and can be obtained for other cities as well. Depending on the authorities, other cities can add or remove indicators unique to the city. Thus, this platform can be replicated and scaled to other cities as well.

5.1 Tool architecture

The Local Economic Intelligence Platform used Jakarta Server Pages for the front end technology and Spring Boot Java, Spring MVC and Spring Security for back-end technologies. Uploading the database was done through PostgreSQL and Data services. The spring boot framework is used for visualization since its framework helps us to create an effective user interface for any web application. User interface of the web application contains the Dashboard, which helps users to navigate through different

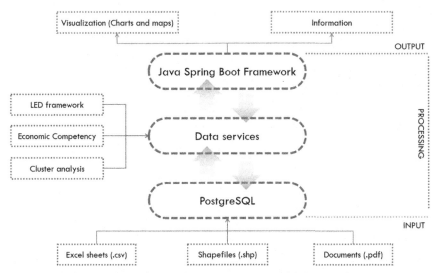

Fig. 1 Representing the tool architecture incorporating the LEIP methodology.

tabs to visualize and understand the local economic development of the city. Graphs are used as a visualization tool in the dashboard so users will have better experience in understanding the visually presented data. The product involved the use of data as a service (Daas) because the data were stored in many places and consumers of the data need ways to find and analyze the information they need without concern for the specific location of that data. Use of "data services" system helps in data security and scalability of this application. Through Data Services, data can be pushed to another web application or mobile application. Local economic development framework is built using predictive analytics Postgresql DB: PostgreSQL, also known as Postgres, which is a free and open-source relational database management system emphasizing extensibility and SQL compliance. PostgreSQL was used in implementation because the application demands the dashboard that helps developers build applications, administrators to protect data integrity and build fault-tolerant environments. The tool architecture is shown in Fig. 1.

Annexure: Indicators used for index building under rapid economic assessment, scoring and data sources

Sr. No.	Theme	Indicator	Unit	Scoring	Data Source
1	Human Capital	District Literacy Rate	%	Higher the better	All India average, KASE report
	Education	Number of Colleges	Number	Higher the better	KASE
		Total number of seats registered	Number	Higher the better	All India urban average
		Total number of research centers	Number	Higher the better	Average
		Total number of students from abroad	Number	Higher the better	Average of South India—Belousova stats
		Total number of vocational training ins	Number	Higher the better	Mode, ASAP, and KASE
		Household expenditure on Education	Number	SD	SD EoL
		Drop-out rate	% :	Lower the better	EoL, KSPB P 315—Kerala Mean
2		National Achievement Survey Score	%	Higher the better	EoL
		Access to digital Education	% :	Lower the better	EoL, all schools = yes (84)
	Employment	Sector-wise employment	%	Higher the better	WFPR—KASE
		Unemployment rate	%	Lower the better	Unemployment average
		Employment growth	%	Higher the better	Census?
3	Investment	Credit Available	Ratio	Higher the better	EoL
	Economic profile	Credit Rating of MC	Class	Higher the better	MC
		Growth rate of district in GVA	%	Higher the better	ER—Kerala
		Number of credit agencies (District)	Number	Higher the better	Industrial profile
4		Amount of Foreign Direct Investment Equity Inflows	Currency (INR)	Higher the better	DPIIT dataset
	Foreign Direct Investment				
5		Procedures for Registering Property (RP)	Number	Lower the better	EoDB—Mean South Asia
	Business profile	Time for RP	Number (Days)	Lower the better	EoDB—Mean SA
		Cost to RP	% of property value	Lower the better	EoDB—Mean SA
		Time for enforcing contract	Number (Days)	Lower the better	EoDB—Mean SA
		Cost to enforce contract	% of claim value	Lower the better	EoDB—Mean SA

#	Category	Indicator	Measurement	Direction	Source
		Recovery rate for insolvency	Number (USD–cents on dollars)	Higher the better	EoDB—Mean SA
		Time for insolvency procedure	Number (Years)	Lower the better	EoDB—Mean SA
		Cost for insolvency procedure	%	Lower the better	EoDB—Mean SA
		Procedures for Construction Permit (CP)	Number	Lower the better	EoDB—Mean SA
		Time for CP	Number (Days)	Lower the better	EoDB—Mean SA
		Cost of CP	% of warehouse value	Lower the better	EoDB—Mean SA
		Paid in minimum capital (Start-up)	% of income per capita	Lower the better	EoDB—Mean SA
6	Infrastructure Road Transport	No: seats available PT	Number	Higher the better	EoL
		Rapid Transport System	Y/N (1/0)	Binary	
		NH connectivity	Number (KM)	Higher the better	LED sheet
		SH	Number (KM)	Higher the better	LED sheet
		Bypass	Y/N (1/0)	Binary	LED sheet
		Ring road	Y/N (1/0)	Binary	LED sheet
7	Air Transport	Distance of airport from ABD	Number (KM)	Lower the better	G-Map (Class)
		Domestic flight destinations	Number	Higher the better	AAI
		International flight destinations	Number	Higher the better	AAI
8	Rail Transport	Grade of RS	Alphabet (Class)	Higher the better	Railways
9	Water Transport	Availability of seaport	Y/N (1/0)	Binary	
10	Supporting Infrastructure	Number of incubation centers	Number	Higher the better	KSUM
		Number of PSUs/LSI	Number	Higher the better	Industrial Profile, Kerala State, (++)
		Number of regional importance institutions	Number	Higher the better	Own compilation (Drive)
		Overall R&D centers	Number	Higher the better	KSIDC dataset

Continued

Sr. No.	Theme	Indicator	Unit	Scoring	Data Source
11	Cluster	Traded Clusters	Number (Score)	Higher the better	EoL
		Cluster Strength	Number (Score)	Higher the better	EoL
12	Digital Infrastructure	Sustained Electricity Interruptions	Number (Score)	Lower the better	EoL
		Number of public wi-fi points	Number	Higher the better	KSITM
		Number of data centers	Number	Higher the better	KSITM + NIC
		Registered Internet Service Providers	Number	Higher the better	
		ICCC	Y/N (1/0)	Binary	
13	Social infrastructure	Number of music/dance/drama theaters	Number	Higher the better	EoL
		Number of restaurants	Number	Higher the better	EoL
		Number of cinema screens	Number	Higher the better	EoL
		Area of open spaces in the city	Number	Higher the better	EoL
		Tourist destinations in the district	Number	Higher the better	EoL
14	Local Policy	Business Implementation Score	State Rank	Number	25/25	KASE
15		Single Window System	Single Window System	Yes/no = 1/0	1	
16		Incentives under Industrial Policy	Tax Incentive/Investment Subsidy	Yes/no = 1/0	1	
		Interest Subsidy	Yes/no = 1/0	1	
		ED/Power Incentive	Yes/no = 1/0	1	
		SD/Registration Fee Incentive—Registration concession	Yes/no = 1/0	0.5	
		Employment generation subsidy	Yes/no = 1/0	0.5	
		Interest free loans	Yes/no = 1/0	0	
		Land Incentives	Yes/no = 1/0	0	

	Land conversion fee concession	Yes/no = 1/0	0
	Entry tax exception	Yes/no = 1/0	0
	Industrial corridor/cluster	Yes/no = 1/0	1
	Capital subsidy	Yes/no = 1/0	1
	Training subsidy	Yes/no = 1/0	0
	Environmental tax reforms/Subsidy	Yes/no = 1/0	1
	Patent registration/Quality certification subsidy	Yes/no = 1/0	0
17	Incentives under IT policy	Land cost rebate linked to employment	Yes/no = 1/0	0
	Power cost incentives	Yes/no = 1/0	0.5
	Incentives related to Stamp Duty, transfer duty and registration fees	Yes/no = 1/0	0.5
	Incentives related to Patent Filling/copyright costs and Quality Certifications Costs	Yes/no = 1/0	0
	Recruitment Assistance Incentives	Yes/no = 1/0	0
	Additional FSI and space utilization of IT parks	Yes/no = 1/0	0.5
	Simplification of labor laws	Yes/no = 1/0	1
	Minimum rates for works contracts for annual maintenance	Yes/no = 1/0	0
	Allowing setting up of IT/ITES in any zone including residential and no-development zones	Yes/no = 1/0	0
	Facilitation for setting up IT units in IT destination, Allotment of land (Bottom-up cluster initiative)	Yes/no = 1/0	1
	Capital subsidy	Yes/no = 1/0	0.5
	VAT/GST Incentive	Yes/no = 1/0	0
	Employment generation subsidy	Yes/no = 1/0	0

Continued

Sr. No.	heme	Indicator	Unit	Scoring	Data Source
		Skill enhancement subsidy	Yes/no = 1/0	0	
		Market Development Assistance	Yes/no = 1/0	0	
		Fiscal Incentives	Yes/no = 1/0	0.5	
		Free and Open Source Software related policies	Yes/no = 1/0	1	
18	Incentives under Start-up Policy	Common facilities center—Warehouse, Storage and Other Physical Common Facilities	Yes/no = 1/0	1	
		Shared services—Legal, Accounting, Patent Filing, Investment Banking	Yes/no = 1/0	1	
		Infrastructure provision—Test lab, Design Studio, Fablabs, Tool Room	Yes/no = 1/0	1	
		Infrastructure incentive—O&M reimbursement, Lease rent exception for a fixed time period	Yes/no = 1/0	1	
		IT Infrastructure	Yes/no = 1/0	1	
		Funding—Seed Fund and Other Funding Initiatives	Yes/no = 1/0	1	
		Financial Assistance as Matching Grants	Yes/no = 1/0	1	
		Infrastructure/Innovation development and maintenance fund	Yes/no = 1/0	1	
		Reimbursement of VAT/CST/GST	Yes/no = 1/0		
			Yes/no = 1/0		

		Reimbursement of Paid Stamp Duty/ Registration Fees	Yes/no = 1/0	1
		Patent filing aid	Yes/no = 1/0	1
		Training assistance	Yes/no = 1/0	1
		Recruitment assistance	Yes/no = 1/0	0
		Internet charge reimbursement	Yes/no = 1/0	0
		Investment Subsidy	Yes/no = 1/0	1
		Promotions incentive: reimbursements of 30% of the actual costs including travel incurred in international marketing through trade shows	Yes/no = 1/0	0
		Performance linked grant for start-ups	Yes/no = 1/0	1
19	Academic connect	Tie-ups with Academic institutions for Start-ups, spin offs aid, Student entrepreneurs		1
20	Engagement with Industries	CIO, Innovation cell initiatives		1
21	Climbing value chain	Policies that encourage firms to set-up research arms, centers of excellence		1
22	MSME Policy	Industrial Promotion Subsidy	Yes/no = 1/0	0
		Interest Subsidy	Yes/no = 1/0	0
		Exemption of Electricity Duty	Yes/no = 1/0	0
		Waiver of Stamp Duty	Yes/no = 1/0	0
		Power Tariff Subsidy	Yes/no = 1/0	0
		Capital incentives/subsidy	Yes/no = 1/0	1
		Quality certification incentives/subsidy	Yes/no = 1/0	0
		Patent registration incentives/subsidy	Yes/no = 1/0	0
		Cost of water audit incentives/subsidy	Yes/no = 1/0	0
		Energy Audit incentives/subsidy	Yes/no = 1/0	0

Continued

Sr. No.	Theme	Indicator	Unit	Scoring	Data Source
		Cost of carrying out credit rating incentives/ subsidy	Yes/no = 1/0	0	
		Acquisition of foreign technology incentives/ subsidy	Yes/no = 1/0	0	
		Market Development Assistance	Yes/no = 1/0	1	
		MSME Cluster Development	Yes/no = 1/0	1	
		Support to sick units	Yes/no = 1/0	1	
		Provision of free tender forms	Yes/no = 1/0	1	
		Loans affordable/subsidy/incentives	Yes/no = 1/0	1	
		Price Preference	Yes/no = 1/0	1	

Acknowledgments

The chapter draws upon field discussions and insights from various institutes of excellence such as the National Institute of Urban Affairs, Centre for Development Studies and Gulati Institute of Finance and Taxation. The project has been immensely enriched by the guidance of Shri Kunal Kumar IAS, Smart Cities Mission Director and Joint Secretary, Ministry of Housing and Urban Affairs. Shri Rahul Kapoor IRAS, Director— Smart Cities Mission, Ministry of Housing and Urban Affairs provided intellectual and operational support throughout the course of the India Smart Cities Fellowship Program. Further, we would like to acknowledge subject-specific contributions of Dr Debjani Ghosh, Senior Research officer, National Institute of Urban Affairs and Dr Amit Kapoor, Honorary Chairperson, Institute for Competitiveness. Shri O P Aggarwal, CEO, World Resources Institute India and Shri Binoy Mascarenhas, Director, UBC Urban Sustainability Initiative, contributed immensely through multiple reviews of Local Economic Intelligence Platform (LEIP). They shared their knowledge and expertise in this niche field and extensively contributed to shaping Local Economic Intelligence Platform (LEIP).

Smt Navjot Khosa IAS, District Collector Thiruvananthapuram and Shri Vinay Goyal IAS, CEO, Smart City Thiruvananthapuram, provided field research support and guided the team with locating appropriate resources. Shri Jitto Jose, a PhD Scholar at Kerala University, brought his expertise in statistical analysis to add value to the methods mentioned in this chapter. The authors take responsibility for any remaining errors in this paper.

Funding

The "Local Economic Intelligence Platform" product development was funded by Smart City Thiruvananthapuram Limited (SCTL).

Disclaimer

This research work is entire that of the author/coauthor and does not carry the views of the organizations associated therein.

References

[1] National Institute of Urban Affairs, Annual Report, National Institute of Urban Affairs, New Delhi, 2019. Available at: https://www.niua.org/sites/default/files/ANNUALREPORT-2018-2019.pdf. (Accessed 5 November 2021).

[2] S. Colenbrander, Cities as Engines of Economic Growth: The Case for Providing Basic Infrastructure and Services in Urban Areas, iied Publications Library, 2016. Available at: https://pubs.iied.org/10801iied. (Accessed 5 November 2021).

[3] R. Mohan, S. Dasgupta, Urban Development in India in the 21st Century: Policies for Accelerating Urban Growth, 2004, Stanford King Center on Global Development Working Paper, Available at: https://siepr.stanford.edu/sites/default/files/publications/231wp.pdf. (Accessed 5 November 2021).

[4] S. Bustos, M. Coscia, M. Yildirim, C. Hidalgo, R. Hausmann, A. Simoes, The Atlas of Economic Complexity, 2014.

[5] R. Hausmann, J. Morales-Arilla, M. Santos, Economic complexity in Panama: assessing opportunities for productive diversification, SSRN Electron. J. (2016).

[6] Global Agenda Council on Competitiveness, The Competitiveness of Cities, World Economic Forum, 2014.

[7] B. Roberts, Tool Kit for Rapid Economic Assessment, Planning, and Development of Cities in Asia, Asian Development Bank, Philippines, 2015.

[8] R. Florida, Regions and universities together can foster a creative economy, Chron. High. Educ. 53 (2016).

[9] S. Lal, H. Gun Wang, U. Deichmann, Infrastructure and City Competitiveness in India, EconStor, 2021. Available at: http://hdl.handle.net/10419/54171. (Accessed 5 November 2021).

[10] A. Shaw, Metropolitan city growth and management in post-liberalized India, Eurasian Geogr. Econ. 53 (1) (2012) 44–62.

[11] A. Mahalingam, PPP experiences in Indian cities: barriers, enablers, and the way forward, J. Constr. Eng. Manag. 136 (4) (2010) 419–429.

[12] JP Morgan Chase and Co, Building Strong Clusters for Strong Urban Economies: Insights for City Leaders from Four Case Studies in the U.S, Competitive Inner City (ICIC), 2017.

CHAPTER FIVE

An investigation into the effectiveness of smart city projects by identifying the framework for measuring performance

Geetha Manoharan[a], Subhashini Durai[b], Gunaseelan Alex Rajesh[c], Abdul Razak[a], Col B.S. Rao[a], and Sunitha Purushottam Ashtikar[a]

[a]School of Business, SR University, Warangal, Telangana, India
[b]GRD Institute of Management, Dr. G.R. Damodaran College of Science, Coimbatore, Tamil Nadu, India
[c]Sri Venkateswara Institute of Information Technology and Management, Coimbatore, Tamil Nadu, India

1. Introduction

Smart Cities are hubs of innovation and economic development. The exchange, optimization, and development of new solutions are unmatched and enormous. However, this transition is proceeding slowly for a variety of reasons. To address issues such as traffic congestion and jamming, energy consumption, resource management, and environmental protection, an innovative approach is required. An ICT-enabled city is often referred to as a smart city. A distinction is made between "intelligent" or "smart" cities. It is a city that has been wired for information technology to be referred to as an "intelligent city." Sustainability is the ultimate goal of a smart city. As a result, performance analysis has become a crucial tool for planning and evaluating city development and projects, as well as assessing cities. City rankings are a hot topic right now, and people are paying more and more attention to them. New residents and investors alike can benefit from comparing cities. Aside from that, it can help cities assess their own strengths and weaknesses and devise development goals and strategies that will help them better fit into the urban system as a whole. Appropriate evaluation mechanisms and indicators are required for proper city comparison, requirement setting during planning phases, and project performance evaluation. This is also true when it comes to implementing smart city strategies and initiatives.

Artificial Intelligence and Machine Learning in Smart City Planning Copyright © 2023 Elsevier Inc.
https://doi.org/10.1016/B978-0-323-99503-0.00004-1

2. Measuring the effectiveness of smart cities

Municipal governments use a wide range of key performance indicators (KPIs) to assess the success of specific projects. These KPIs could represent the city's environmental, social, and economic objectives. Goal-oriented metrics that are consistent with sustainability principles can guide cities in developing indicators to measure progress toward the goals in each smart city characteristic. In general, metrics should be tied to the company's vision and goals. A number of initiatives are currently underway by cities, businesses, research organizations, and government agencies to assess a city's sustainability or environmental impact and smart city initiatives. It has recently become possible to assess ICT products, networks, and services' environmental impact by applying life cycle thinking. As part of various initiatives, many cities must also report their greenhouse gas emissions, greenhouse gas emissions, and energy consumption. Frameworks for assessing a smart city often rely on a classification system for people (society), planet (environment), and profit (economy). Assessment frameworks typically include indicators for the economy, people and quality of life, and governance or services. Most frameworks also include environmental indicators that consider things like energy consumption, building sustainability, carbon footprint and waste production. The systems also address smart mobility and transportation, but the emphasis varies depending on the framework. The role of ICT as an enabler technology is deeply embedded in all of the major subfields. However, there have been no frameworks focusing on the scalability and replication of smart city solutions to date.

3. About measurement concept

Over the last two centuries, the triple bottom line of social (People), environmental (Planet), and economic (Prosperity) sustainability has been widely accepted in the development of indicator systems for national and regional urban development. Corporations are increasingly focusing on the three Ps in their reporting (people, planet, and prosperity). Smart city projects are assessed based on their potential impact on a number of social and economic metrics. However, this is insufficient to ensure the success of a smart city initiative. A project's success is also determined by how it has been or will be implemented in a variety of contexts. People, planet, and prosperity (PPP) are all indicators of a city's well-being. Because the quality of the city context (external factors) and development and implementation processes must be assessed, a number of indicators are required (internal factors). These projects' ultimate impact on energy and CO_2 emission reduction goals are determined by their ability to be replicated in other cities.

The reasons for designing frameworks using indicators are such as:

➢ Policy goals and key indicators should be linked to show how progress is being made toward these goals.

➢ Allowing individuals or groups of individuals to assign weightings to policy objectives and use them as a benchmark for comparing indicators (and thereby to the indicators belonging to a subtheme).

➢ To make it easier to communicate the results of the indicators to those who make decisions.

➢ Performance measurement systems are needed by cities in order to evaluate strategies and projects, and to determine how a particular project contributes to a city's long-term strategy.

➢ An adaptable, user-friendly, secure, and compatible system should be implemented.

4. Performance measurement and its dimensions

Certain data must be available and reliable in order for a framework to calculate several indicators. The project's data collection and the city indicators' data collection are two separate processes. The smart city project indicators are designed to assist in assessing the project's success and replication potential. Through the use of both quantitative and qualitative indicators, this can be achieved. It is necessary to collect qualitative data by interviewing stakeholders or by analyzing policy documents. Data extraction from the project documentation is required for quantitative indicators. While streamlined data collection is preferable for many smart city projects, some indicators necessitate qualitative data that can only be obtained by directly participating in the project (e.g., through interviews, questionnaires).

A large number of city indicators can be obtained from statistical sources both within and outside the city administration. In some cases, that data are presented in the form of year-to-year averages for the entire city. The distinctions between districts in a city may be more interesting. However, using geospatial data, indicators can be calculated for constrained areas like city districts.

The availability, sources, and reliability of relevant datasets were examined, including data access methods, existing formats, and the level of confidentiality. Each participating city's availability rates for the datasets required by the smart city KPIs were assessed. Quantitative or qualitative measures must be used to evaluate a project's KPIs, which mean that project managers or other stakeholders must be interviewed to gather this information. Additionally, each project must define its own set of data boundaries, as the scope of relevant data is always project-dependent. Not all indicators will be available immediately in every city. In the beginning, a city's use of smart city indicators sets off a chain reaction. The development of city-specific indicators and data collection mechanisms is essential for cities.

This means that there is a wide variation in how different countries, cities, and city departments define the same datasets. The quality of the overall assessment is dependent on the quality of the indicators, which in turn is dependent on the underlying data. As a result, maintaining high levels of assessment and comparability requires careful management of data quality throughout the process. The reliability of the data and the results of the corresponding indicator can be explained in part by communicating transparently all metadata that underlie the datasets when comparisons are conducted. Because of privacy and confidentiality concerns, some data cannot be released in its raw form to the general public. In accordance with privacy regulations, cities may keep these data in their internal systems if they follow the access rules and conditions specified in their privacy protocols. Additionally, a prototype web-based performance measurement tool was developed, which combines data entry, calculation methods and result visualization in an intuitive, user-friendly interface.

This is advantageous when cities make updates to their datasets. The user interface allows the user to enter new KPI values, view previously entered values, and download previously entered KPI data in Excel spreadsheets. There is a list of all the key indicators on the indicator selection user interface, along with the most recently entered value and a link to the value input page, along with general information about the city being studied. You can also navigate between indicator subcategories. The user interface for entering KPI values includes a complete description of the project and allows the user to enter the assessment time and date of the data, the assessment value, the indicator performance level on a 1–5 scale, and additional information such as data sources, comments, and so on.

Additional features include comparisons between assessments conducted at different points in time and visualization of indicators by assessment time.

Trend graphs can be used for quantitative and performance level (Likert) indicators. Indicators that are automatically read and others that have multiple times stamped values stored in the system are particularly relevant to this (e.g., air quality index and energy or water consumption-related indicators). Due to the framework's use and the improvement of some KPI descriptions, recommendations have been made. Cities provided feedback and suggestions on the usability and functionality of early prototypes of the KPI tool during the co-design process.

Certain studies concentrate on the measurement dimensions and their impact on the industry's outcomes. According to Crowther [1], business performance has three dimensions: (i) perspective, (ii) goal, and (iii) focal. Forker et al. [2] investigate how quality affects a business's performance. Hedges and Moss [3] discuss the challenges associated with determining the cost-effectiveness of training programs. Using a result-oriented approach to system improvement, Hicks [4] proposed.

Quality, delivery, customer satisfaction, cost, and security are the six categories in which key performance factors are introduced [5]. Lofsten [6] presents a model of partial productivity in which production changes over time. Mapes et al. [7] identify the factors that lead to high productivity, quick delivery, and consistent quality. Finally, some studies also identify that some processes are needed to be aligned with the performance measurement strategy [8]. Performance measurement for a smart city management can be performed using different dimensions. Some of the important dimensions include productivity, quality, timely delivery, internal control system, and external factors. In terms of operational performance, productivity can be defined as the relationship between input and output [9]. Productivity is viewed as physical quantity, while fixed capital or working capital inputs are viewed as capital productivity. To measure productivity, direct labor is measured mainly in terms of total productivity, labor productivity, indirect labor, and productivity of fixed capital and working capital.

The operating result can be conceptually divided into two areas: costs, including production costs, and productivity [10]. The costs of production are differentiated by a direct link, which can be explained by mathematical formulas, with the final results of the company, namely net profit and profitability; Economic design in terms of time, flexibility, and quality. Quality-related yields [11] are also distinguished by quality, perceived quality, quality of care, and quality costs.

The time aspects are divided into external and internal. The internal time gains are divided as waiting and shifting the times of the other [12].

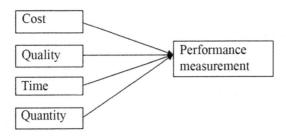

Fig. 1 Performance measurement and its dimensions.

Perceived time performance from the outside is subdivided into system hours (including delivery, manufacturing, and distribution conditions); Speed and reliability of deliveries (suppliers and customers) and time to market (or time spent developing a new product). As a result, the performance measurement literature addresses various dimensions such as cost, time, quality, quantity, profitability, flexibility, productivity, as well as internal and external factors related to the manufacturing process of the organization (Fig. 1).

5. Performance effectiveness and its dimensions: Product measures

Many studies have also suggested many factors that contribute toward measuring organizational effectiveness, especially in the operating type of industry [13]. One such important factor is "product." Traditionally, organizations were majorly product-focused [14]. During those times, organizational effectiveness was measured based on the capacity of the firm to produce a product in terms of volume. Sometimes, the product mix was also done to increase the effectiveness of the firm [15]. In recent years, firms are more focused on the flexibility of the firm to meet out their demand based on volume and quality [16]. Having customer satisfaction in mind, these firms started producing products based on customer specifications and requirements. Thus, the innovation of the firm played a vital role in satisfying their customers by producing the product based on their specifications. The product quality is also considered to be one of the important aspects to measure the effectiveness [17].

5.1 Process measures

The process focus is suitable for firms having complex or divisible processes [18]. Process focused firm concentrates on reduction in cost of production and other related costs. It smoothens the process in product mix that leads to effectiveness of the organization [19]. For operating type of firms, analyzing the capacity, technology used in production, maintenance of documentation of each activity, effective communication of the functions, and all the other related processes that support the manufacturing activities are to be measured [10]. The quality management system is an important process in an operating type of firms [20]. The activities like timely service and delivery also contribute toward effective internal process in an organization. In recent years where human resources are considered to be the valuable asset for an organization, the effectiveness of training given to the employees and introduction of development programs also contributes toward the overall effectiveness of the organization. People-related process was also considered to be one the major factors to increase the organizational effectiveness, since skilled labors and experts in the field of production are considered to the backbone for these types of industries.

5.2 People measures

Organizational effectiveness is based on the people who are doing the business in and around the firm [21]. People who are involved in a firm are termed as stakeholders, namely, employees, shareholders, suppliers, customers, intermediate agents, government agencies, competitors, and the public [22]. Most of the studies show that firms are more concerned with the customer and obtain feedback on their service and product regularly to maintain or increase the level of customer satisfaction. Several studies also suggest developing the skills of the employees by way of training and development programs. The level of employee satisfaction is directly related to the level of employee motivation. Motivation by way of remuneration or recognition is practiced in firms to motivate their employees [23]. Of all these activities, the most important and considerable factor is communication. It impacts the decisions taken by the shareholders. Only effective communication between the people working in any organization will help to avoid unnecessary deviations from the goals and support effective decision-making [24].

5.3 Policy measures

As per the views of most of the industrialist and researchers, the policy of an organization is considered to be very important to increase its goodwill and the best way to become market leaders [25]. Generally speaking, the policy of any organization includes their ethics in business, their dedication toward following statutory regulations, and other standards both internal and external, nature conservation, CSR activities for employees and society, etc. [26].

Some industries have policies framed to become financially sound; to be a more profitable company and to get certified like ISO, TS certification. Industrialists are also more concerned in framing policies for internal activities that support the production process like reducing cost on energy and water consumption, regular inspection of tools and types of equipment, maintenance of documentation at all levels of activities carried on in the firm, etc. [27]. Some of the studies say that attracting new customers is considered an important policy of manufacturing firms. In recent days, green practices and sustainable practices are also found a place in a firm's policy.

5.4 Place measures

Here, "place" refers to the workplace or the environment of the manufacturing units. When it comes to the internal workplace, any firms must have effective safety measures and practices [28]. Since these types of industries have to deal with various types of equipment, tools, machinery, chemicals, processes, and related supporting activities, the firm should make it compulsory to follow the standard safety measures to safeguard their internal environment [29]. When it comes to the external environment, firms are considered to be more responsible only when they follow environmental safety measures and nature protection measures. Most of the firms are going green to protect the environment from getting polluted by the wastes produced from their industry. Thus, protecting and safeguarding the people and the environment increases the effectiveness of the organization. Based on the above discussion, the effectiveness of the 5Ps will lead to the overall effectiveness of the organization. Thus, based on the above discussions, the following 5P framework can be formulated (Fig. 2).

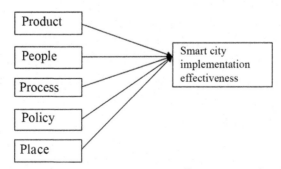

Fig. 2 5P Framework: Smart city implementation effectiveness and its dimensions.

6. Measurement models used in industries

Various approaches to measuring organizational effectiveness lead to the identification of four basic elements or components of organizational effectiveness. These include the human relations model, which emphasizes flexibility and internal focus and also highlights the criteria for achieving the goal [30]. And it also focuses on the role of information management and communication. The Rational Goals Model places a strong emphasis on external priorities, planning, goal setting, productivity, and effectiveness. The open system model focuses on flexibility and external orientation. It also focuses on preparation, growth, resource acquisition, and external support. When it comes to comparing companies using a single set of criteria, the framework assumes that there is no better point of comparison (e.g., efficiency, flexibility, and return on investment). An organization's effectiveness is subjective and based on the organization's values and objectives. Kaplan and Norton's [31] Balanced Scorecard and EFQM Business Excellence Model are the two most popular performance measurement systems [32]. Both offer a structured approach to identify opportunities for improvement and threats and translate the company's strategy into achievable goals and specific tasks. Unlike these systems, the Performance Matrix [33], the SMART Performance Pyramid [34], the Performance Prism [35], and Kanji Business Excellency Performance System [36] are also developed but not used predominantly. Beischel and Smith [37] provide a framework for measuring performance. McGrath and Romeri [38] present the R&D Efficiency Index to measure the overall success of product development efforts. Some studies present about assessing the level of measurement of quality and quality performance results [39]. Motwani et al. [40] present a model that helps entrepreneurs control their productivity. As expected, each performance measurement model developed should be based on the success factors of each firm [41]. It should be implemented in the strategy and monitor the performance of the company. One of the most important aspects is that it must be integrated into a strategic system based on organizational objectives, critical success factors, and financial and nonfinancial aspects. Researchers, therefore, expect the performance measurement system to evolve dynamically with strategy, in which it must be associated with reward systems [42].

7. Designing a framework for smart city performance measurement

An organization's performance can also be evaluated using critical success factors (CSFs) [43]. CSFs are factors related to the vital problems of the organization and future success. Some studies state that recruitment efforts are focused on CSFs, with emphasis on only certain aspects of an organization rather than the entire organization [44]. It is found that in assessing performance, it was important to identify critical factors for the success of the organization [45]. Also, these factors (CSFs) show signs of an effort to achieve the organizational goals which include helping the administration to understand, manage, and evaluate the progress of the organization [46]. Key success factors (KSFs) and critical success factors (CSFs) must be identified and understood by organizations in order to achieve their mission and goals [47]. As a term of reference for data analysis and business intelligence, the term CSF was first used. For example, user participation is a critical success factor for a successful IT project. It is important to distinguish between critical success factors and success criteria. Because they are the fruits of an organization's labors, they must be deemed fruitful. Key performance indicators (KPIs) can be used to measure the success of a project [48]. It is unique to each organization and will reflect current activities and future goals.

To measure each of the CSF, required strategic objectives can be assigned. These strategic objectives are very important to be assigned because it helps to know whether the CSF has been achieved or not. For every strategic objective, it is important to have a corresponding key performance indicators and metrics. These KPI and metrics can be calculated to know and understand the status of the CSF that has to be achieved. Finally, these metrics can be grouped as cost metrics, time metrics, quality metrics, and quantity metrics [49]. Based on the abovementioned discussion, the following framework has been designed to measure the performance effectiveness of smart cities (Fig. 3).

8. Conclusion

KPI frameworks and associated tools may be used in the future by cities and policymakers. Smart city project consortiums involved in R&D projects are also expected to utilize this framework for assessing impact. These

CSF: Critical Success Factor

KPI: Key Performance Indicator

SO: Strategic Objectives

Fig. 3 Framework for smart city performance measurement.

kinds of projects necessitate a combination of standard input or output reporting (e.g., the number of apps or sensors implemented in a project) with impact reporting. KPI and metric assessment have been incorporated into some lighthouse project consortiums, but this is not the case for all. From the extensive testing done during the development of KPI frameworks, the following summarizes the most important findings:

This project would not have been possible if cities had not actively participated throughout its development (both the KPI framework and performance measurement tool). In addition to the user interface, the ability to read data automatically should be tested. In addition, cities can use the tool's APIs to connect their own datasets to it. As a result of numerous case studies, the majority of the project KPIs has been proven to be accurate. It is possible to implement a wide range of KPI improvements. Assessment methodology for smart city projects has been proven to be useful in practice and thanks to its well-received balance of qualitative versus quantitative KPIs, which are aligned with the strategic goals of smart city projects. The KPI framework can be used for multiple purposes, but it has been found that a flexible approach based on city or case-specific objectives and priorities is appropriate. While still allowing for a comprehensive and holistic assessment, this allows for the evaluation of only the most relevant KPIs and not a lack of information or expertise, but a problem with where and how the information is located and accessible. As a result of the difficulty of locating and evaluating KPIs, many cities have given up on the KPI evaluation process entirely. An open data management, storage, and publishing platform would greatly assist in the localization and exploitation of the enormous amount of city data that are currently out there. The next step would be to standardize (open) data set formats, which would improve data exploitation even more. Aside from boosting efficiency in city operations, these steps would also improve management and coordination of smart city activities. To integrate open datasets from different URLs in the future projects, linked data can be used as it is a good option.

References

[1] D.E.A. Crowther, Corporate performance operates in three dimensions, Manag. Audit. J. 11 (8) (1996) 4–13, https://doi.org/10.1108/02686909610131639.

[2] L.B. Forker, S.K. Vickery, C.L.M. Droge, et al., The contribution of quality to business performance, Int. J. Oper. Prod. Manag. 16 (8) (1996) 44–62, https://doi.org/10.1108/01443579610125778.

[3] P. Hedges, D. Moss, Costing effectiveness of training: case study 1— improving parcel force driver performance, Ind. Commer. Train. 28 (3) (1996) 14–18.

[4] D. Peterson, M.D. Hicks, Leader as Coach: Strategies for Coaching and Developing Others, Personnel Decisions International Corporation, Minneapolis, 1996.

[5] S. Mathiyalakan, C. Chung, A DEA approach for evaluating quality circles, Benchmark Qual. Manag. Technol. 3 (1996) 59–70.

[6] H. Lofsten, Measuring maintenance performance—In search for a maintenance productivity index, Int. J. Prod. Econ. 63 (1) (2000) 47–58.

[7] J. Mapes, M. Szwejczewski, C. New, et al., Process variability and its effect on plant performance, Int. J. Oper. Prod. Manag. 20 (7) (2000) 792–808.

[8] M. Bourne, M. Kennerley, M. Franco-santos, Managing through measures: a study of impact on performance, J. Manuf. Technol. Manag. 16 (2005) 373–395.

[9] J. Sauermann, Performance Measures and Worker Productivity, IZA World of Labor, 2016, pp. 1–11, https://doi.org/10.15185/izawol.260.

[10] R.S. Kaplan, Productivity Measurement and Management Accounting, 1986.

[11] I. Chen, 'Critical quality indicators of higher education, Total Qual. Manag. Bus. Excell. 28 (2017) 130–146.

[12] J. Singh, G. Singh, H. Singh, A Framework for Measuring on-Time Delivery Performance in Processing Organization: A Case Study, 2019.

[13] Y.-T. Cheng, H.-H. Chou, C.-H. Cheng, et al., Extracting key performance indicators (KPIs) newproduct development using mind map and Decision-Making Trial and Evaluation Laboratory (DEMATEL) methods, Afr. J. Bus. Manag. 5 (26) (2011) 10734–10746..

[14] M. Madaleno, C.A. Varum, I. Horta, SMEs performance and internationalization: a traditional industry approach, Ann. Econ. Financ. 192 (2018) 605–624.

[15] L.M. Chuang, A reconceptualization of manufacturers' sustainable product-service business models: triple bottom line perspective, Adv. Manag. Appl. Econ. 91 (2019) 1792–7552. Available from: http://www.scienpress.com/Upload/AMAE%2FVol9_1_4.pdf.

[16] S.A. Carlsson, D.E. Leidner, J.J. Elam, Individual and organizational effectiveness: perspectives on the impact of ESS in multinational organizations, in: Implementing Systems for Supporting Management Decisions, 1996, pp. 91–107, https://doi.org/10.1007/978-0-387-34967-1_7.

[17] G. Dalmaso, Quality Concept, 2012.

[18] M.T. Sweeney, M. Szwejczewski, Manufacturing strategy and performance, Int. J. Oper. Prod. Manag. 16 (1996), https://doi.org/10.1108/01443579610113924.

[19] C.E. Beck, G.R. Schornack, Management and Organizational Processes, 2001.

[20] A.J. Lordsleem, C. Duarte, B.J. Barkokébas, Performance Indicators of the Companies Quality Management Systems With ISO 9001 Certification, 2010, pp. 567–578.

[21] A. Andrew, Employees' commitment and its impact on organizational performance, Asian J. Econ. Busin. Account. 52 (2017) 1–13, https://doi.org/10.9734/ajeba/2017/38396.

[22] J.B. Harer, B.R. Cole, The importance of the stakeholder in performance measurement: critical processes and performance measures for assessing and improving academic library services and programs, Coll. Res. Libr. 662 (2005) 149–170, https://doi.org/10.5860/crl.66.2.149.

[23] I. Gabcanova, Human resources key performance indicators, J. Compet. 41 (2012) 117–128, https://doi.org/10.7441/joc.2012.01.09.

[24] F. Bergeron, C. Bfigin, The Use of Critical Success Factors in Evaluation of Information Systems: A Case Study, 1989, p. 54.

[25] D. Vazquez-Bustelo, L. Avella, The effectiveness of high-involvement work practices in manufacturing firms: does context matter? J. Manag. Organ. 252 (2019) 303–330, https://doi.org/10.1017/jmo.2016.69.

[26] H.C. Katz, T.A. Kochan, R. Weber, Assessing the effects of industrial relations systems and efforts to improve the quality of working life on organizational effectiveness, Acad. Manag. J. 28 (1985) 509–526.

[27] D.R. Dalton, W.D. Todor, G.J. Fielding, L.W. Porter, Organization structure and performance: a critical review, Acad. Manag. Rev. (1980) 51.

[28] K. Ghorbannejad Estalaki, On the impact of organizational structure on organizational efficiency in industrial units: industrial units of Kerman and Hormozgan provinces, UNIFAP 73 (2017) 95, https://doi.org/10.18468/estcien.2017v7n3.p95-105.

[29] C.F. Gomes, M.M. Yasin, J.V. Lisboa, A literature review of manufacturing performance measures and measurement in an organizational context: a framework and direction for future research, J. Manuf. Technol. Manag. 156 (2004) 511–530, https://doi.org/10.1108/17410380410547906.

[30] A. Shahin, M.A. Mahbod, Prioritization of key performance indicators: an integration of analytical hierarchy process and goal setting, Int. J. Product. Perform. Manag. 563 (2007) 226–240, https://doi.org/10.1108/17410400710731437.

[31] R.S. Kaplan, D.P. Norton, The Balanced Scorecard: Translating Strategy into Action, Harvard Business School Press, Boston, MA, 1996.

[32] U. Nabitz, N.S. Klazinga, A. Medisch, C. Universiteit, J. Walburg, The EFQM Excellence Model: European and Dutch Experiences With the EFQM Approach in Health Care, Available from:, 2000, https://doi.org/10.1093/intqhc/12.3.191.

[33] S. Lavy, J.A. Garcia, M.K. Dixit, Developing a categorization matrix of key performance indicators KPIs: a literature review, in: International Conference in Facilities Management, 2010, pp. 407–419.

[34] P. Everett, SMART Goals: A How to Guide, 2016, pp. 1–13. Available from: https://www.ucop.edu/local-human-resources/_files/performance-appraisal/HowtowriteSMARTGoalsv2.pdf.

[35] A. Neely, C. Adams, A. Consulting, Perspectives on Performance: The Performance Prism, 2001.

[36] G. Kanji, Measuring business excellence, Routledge Adv. Manag. Bus. Stud. (2002).

[37] M.E. Beischel, K.R. Smith, Linking the shop floor to the top floor: here's a framework for measuring manufacturing performance, Manag. Account. (US) (1991) 9–25.

[38] M.E. McGrath, M.N. Romeri, The R&D effectiveness index: a metric for product development rerformance, J. Prod. Innov. Manag. 11 (4) (1994) 213–220.

[39] N. Hietschold, R. Reinhardt, S. Gurtner, Measuring critical success factors of TQM implementation successfully—a systematic literature review, Int. J. Prod. Res. 52 (2014) 6254–6272.

[40] J. Motwani, A. Kumar, M. Youssef, Business Process Reengineering: A Theoretical Framework and an Integrated Model, 1998, https://doi.org/10.1108/EUM0000000004536.

[41] R.A. Dickinson, Critical Success Factors and Small Business, 1984, pp. 49–58.

[42] J.O. Kuykendall, Key Factors Affecting Labor Productivity in the Construction, Master Thesis, University of Florida, 2007.

[43] U. Silesia, C.M. Olszak, E. Ziemba, Critical Success Factors for Implementing Business Intelligence Systems in Small and Medium Enterprises, 2012.

[44] A. Jenko, M. Roblek, A primary human critical success factors model for the ERP system implementation, Organizacija 49 (2016) 145–160, https://doi.org/10.1515/orga-2016-0014.

[45] J.L. John Latham, M. Hollis, J. Whitlock, in: U.K.J. Latham (Ed.), Comprehensive Strategic Performance Measures: The Key Success Factors and Associated Measures Employed by Senior Executives to develop and Implement Future Strategies, 2013.

[46] M. Saayman, C.A. Klaibor, Critical success factors for the management of 4x4 ecotrails, South Afr. J. Busin. Manag. 47 (2016) 45–55.

[47] H. Barrett, E. City, Success factors for organizational performance: comparing business services, in: Health Care and Education, 2002, pp. 16–29.

[48] S. Lavy, J.A. Garcia, M.K. Dixit, Establishment of KPIs for facility performance measurement: review of literature, Facilities 28 (2010) 440–464, https://doi.org/10.1108/02632771011057189.

[49] D. Subhashini, R. Krishnaveni, Designing a Leaf-Based Performance Metric Model for Organizational Effectiveness (PMMOE) With Special Reference to Small and Medium Enterprises (SMEs), 2020. Master Thesis.

PART TWO

Smart water management

Waste water-based pico-hydro power for automatic street light control through IOT-based sensors in smart cities: A pecuniary assessment

Tapas Ch. Singh[a], K. Srinivas Rao[a], P. Srinath Rajesh[b], and G.R.K.D. Satya Prasad[a]

[a]Department of Electrical and Electronics Engineering, GIET University, Gunupur, Odisha, India
[b]Department of EEE, Abdul Kalam Institute of Technological Sciences, Kothagudem, Telangana, India

Abbreviations

PHP	pico-hydro power plant
SW	sewage water
IOT	Internet of things
LED	light emitting diode
BES	battery energy storage
UPS	uninterrupted power supply
LDR	sensors light-dependent resistor
WtE	waste to energy
CCTV	closed-circuit television

1. Introduction

More or less, every moment we are in the verge of adapting to a novel thought of how we can best make use of innovation in sustainable and efficient power generation. A smart city employs new creative methods to fortify available resources for the better use of societal needs by smart communication, smart technology, smart monitoring, and smart service for future growth [1]. Resource supervision and utilization is vital for smart city, and proper monitoring through data analysis, data processing, and data security is necessary for its application [2]. More than 50% of the whole population live in cities as smart technologies and urban infrastructure attract people more toward smart cities [3]. The use of devices having frequent access to the internet for proper communication and management is most common with millions of people [4]. The smart city facilitates various systems such as channelizing proper drainage system, traffic management, smart street light, air quality monitoring, uninterrupted power supply (UPS), waste management, electrical energy management, etc., most efficiently, hence enhancing the quality of life in smart cities.

In a time of energy crisis and inadequate energy management, there is a need of proper energy utilization and generation with reduced loss. Distribute generation system by considering various local hybrid generation resources such as solar, wind, hydro, battery, and generator power is a most common feature of developing cities. In this paper, a PHP-based energy storage scheme is considered for power generation for street light operation, traffic light control, and CCTV surveillance in city. Pico-hydro turbines (impulse/cross flow) are more efficient for low-head applications. Form the theoretical and practical observations, it is found that a water flow rate of 85 l/s will yield a maximum up to 5 kW of power [5]. Such generated power is enough for the minimum utilization of number of low-power high lumen LED for street lights [6]. The conducted experiment for power generation through pico-hydro plant which used a proper overhead tank of 0.4 million capacity for a maximum discharge of 9 l/s of water from a net head of 12 m generates nearly 0.5 kW of power. The amount of water discharge per person per day is 81 gallon and average number of persons per home is considered as three [6]. Hence, a fixed amount of 245 gallon per home is estimated for the proposed system. The overall system working methodology is depicted in Fig. 1.

The Internet of Things (IoT) is an internet working of physical objects which enables users to interact with one another. The Internet of Things (IoT) enables remotely sensed management of objects. It is a sophisticated mechanism and monitoring system that employs human intelligence to provide improved and controlled items/solutions. Such technologies provide more accessibility, management, and efficiency [7]. IoT offers a variety of industrial applications, including smart homes, parking management, smart highways, and smart lighting, among others. The present manual street lighting method has a lot of drawbacks, including upkeep, scheduling, and network congestion [8]. IoT technology potentially helps solving these issues. The system relies on autonomous street lighting and operation that is sophisticated and environment-adaptive [9,10].

Power consumption by the street light is reduced by the smart monitoring of working hours through IOT-based device which gives a new path to the energy management system. NodeMCU-based ESP8266 Wi-Fi module, ultrasonic, and LDR sensor are used. LDR receives the day light intensity level based on which ESP8266 is used for switching operation. Ultrasonic

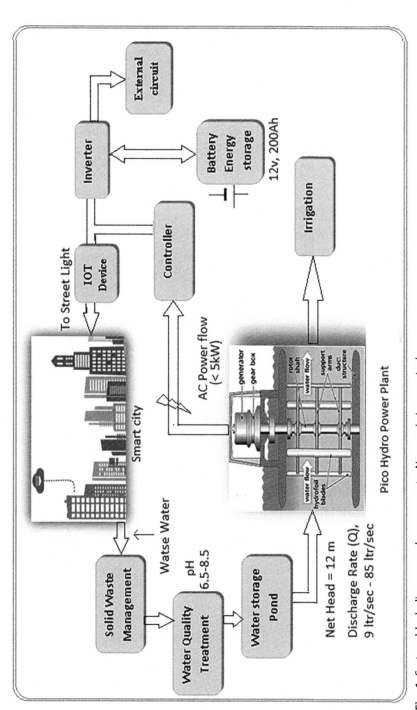

Fig. 1 System block diagram and components. *No permission required*

sensor helps in power saving through minimal operation of lights except during passage of any object below the streetlight. Intensity of light increases from 60% to 100% when moving object is detected by the ultrasonic sensor that shaves nearly 40% of the power consumption by the street lights [11]. Yanka and Baskaran [10] shows an adaptive street lighting system by using LDR, IR sensor, and microprocessor to monitor ON–OFF of lights. A systematic control of switching streetlight from certain distance before entering into street light zone is proposed by connecting series of sensors, placed apart from each other to detect the motion of the object [3].

2. System description

2.1 Waste water collection analysis

Water, which is releasing from the residential area, is contaminating and creating environmental pollution in general around them. WtE management is the main apprehension of the proposed system for the development of a smart and clean ecological sustainable future. From Homewateruse [12], it is considered that the average water used by an individual person per day is more than 81 gallons as shown in Table 1. For a smart city having a capacity of 500 to 5000 numbers of houses, the average amount of wastewater discharges is more than 700 gallons per day (for 3 persons/house) from an individual house as shown in the data mentioned in Table 2. For a typical smart city, electrical power generation through wastewater is nearly equal to the power consumption by the street lights in normal operation, hence experiment is done by considering minimum water availability per day.

2.2 Sewage water treatment

Raw sewage water passes through the screening chamber where physical separation of floating solid, organic, and inorganic materials takes place. Then coarse particles like sand, ash, inert materials, etc., are removed

Table 1 The average water utilization by an individual person per day.

Items description	Wastewater in gallons
Full tub	36
Flush	20
Showers	20
Washing machine	15
Dishwasher	8
Hygiene	2.5
Total	81.5

Table 2 Average amount of wastewater discharges from an individual house per day.

Sl. no.	Water discharge rate (Gallon/s)	Amount of water discharge to turbine in 12 h (Gallon)	Average water available per home (Gallon)	No of available homes (Approx.)	Total availability of water (Gallon)
1	2.3	27.6	245	500	122250
2	5	60	245	1000	245000
3	8	96	245	1500	367500
4	11	132	245	2000	490000
5	13	156	245	2500	612500
6	16	192	245	3000	735000
7	18.5	222	245	3500	857500
8	21	252	245	4000	980000
9	22.5	270	245	4500	1102500

through grit removal process which is followed by a primary classifier that makes the water suitable for use in the turbine as shown in Fig. 2.

Sewage water is treated through various processes to make it suitable for turbine operation. Because the untreated water is having more pH value and suspended solid levels which would decrease the turbine efficiency and the turbine material may get corroded. Hence, it leads to more turbulence and clogging of the turbine. Table 3 shows the allowable result of treated water for the proposed system.

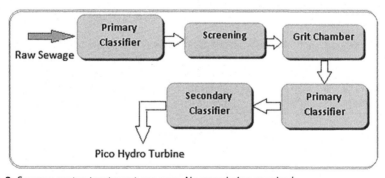

Fig. 2 Sewage water treatment process. *No permission required.*

Table 3 Allowable parameter values of treated water plant.

Parameter	Value
pH	6.5–8.5
Total suspended solids	<100 milligram/liter
Total dissolved solids	<1800 milligram/liter

2.3 Pico-hydropower generation

The proposed system is considered with a water flow rate of 9 liters per second with a gross head of 14 m starting as a base point. The net head offered with water is measured as 12 meter wherever a height of 2 meters less is considered as a loss that occurred due to friction in the pipe connecting between the water reservoir and generator set at the lower base. The time of operation of this system is only during the night time for the working of street light in the city. Hence 12 hours per day is considered for the theoretical power calculation as mentioned below [6]. The details of electricity generation of different flow rate and amount of revenue generated per year are mentioned in Table 4 where the approximate number of LED street lights is calculated through an analytical method for smart city application.

The return on investment of the models is used to assess the viability of the suggested technology. The accounting rate of return is the amount of time it takes to recoup the cost of acquiring and implementing the equipment. The generator turbine couple, water turbine, and reservoir for the wastewater storage are the major components involved. The cost of the installation, labor cost, and overall alignment are all included in the overall cost of the projected structure in use for payback calculation.

Gross head offered $(H) = 14 \, \text{m}$

Net obtainable head $(Hn) = 12 \, \text{m}$

Water flow rate $(Q) = 9 \, \text{l/s}$

Table 4 Electricity generation for different flow rate and revenue generated per year.

Sl. no.	Hours of running	Electrical power generated in (kW)	Units of energy generated for 12 h (kWh)	Revenue generated in (Rs) per year	Approx. no of LED bulbs (30w)
1	122250 / (2.3 x 3600) = 14.76	0.50	6.00	10,950	17
2	13.61	1.00	12.00	21,900	34
3	12.76	1.50	18.00	32,850	50
4	12.37	2.10	25.20	45,990	70
5	13.09	2.50	30.00	54,750	84
6	12.76	3.10	37.20	67,890	104
7	12.88	3.60	43.20	78,840	120
8	12.96	4.10	49.20	89,790	137
9	13.61	4.50	54.00	98,550	150

Working period $(T) = 12\,\text{h}$

$$\text{Hydraulic power } (P_H) = \text{Head (in m)} \times \text{Flow (in l/s)}$$
$$\times \text{Turbine efficiency} \times 9.81 \qquad (1)$$

$$= Hn \times Q \times \eta \times g = 12 \times 9 \times 0.7 \times 9.81 = 741.636\,\text{W}$$

Mechanical power = Net hydraulic power × Mechanical efficiency $\quad$ (2)

$$= 741.636 \times 0.9 = 667.472\,\text{W}$$

Electrical power = Mechanical power × Generator efficiency $\quad$ (3)

$$= 667.472 \times 0.7 = 467.23\,\text{W}$$
$$= 0.5\,\text{kW (Approx)}$$

Quantity of energy formed $= 0.5 \times 12$ (assume 12 h per day) $\quad$ (4)

Total energy formed per day $= 6\,\text{kWh}$
Cost of electricity/day = Rs. 5 per unit
Cost saved/year $= 6 \times 5 \times 365$ (Assuming 365 days of process) = Rs.10,950/–

Payback period = Project Cost/yearly savings $\qquad$ (5)

$$= 60000/10950 = 5.47 \text{ years}$$

Therefore, from the above hypothetical calculation, it shows that the minimum yearly saving is Rs. 10,950 that leads to the payback period of less than 6 years.

2.4 BES system design and analysis

Battery inverter scheme is very much common in residential application and workplaces where they are used to provide uninterruptible power to the connected load. A proper inverter sizing is required to meet the expected load demand so that it can work normally. Inverter sizing is done by using the available load that is directly connected to Pico-hydro system. Approximately 3.6 Kw of the lighting load is considered and ensured continuous working through the UPS in case of failure of Pico-Hydro system.

The primary objective of the BES system is to charge during surplus power from Pico-hydro and discharge during pico-hydro breakdown or the stored energy can be utilized for the traffic light control and CCTV monitoring of smart city during day time. A proper inverter design is carried out and approximately 5 KVA inverter is desirable for the required load demand, and the rating can be decreased when the load will be adjusted with minimum load; mathematically the power available by the inverter is calculated as follows.

Inverter sizing:
The total available light load is 0.5 kW

$$\text{Power (kVA)} = \text{Power in kW/PF} \tag{6}$$

= Power in kW/0.8 (nominal PF = 0.8, which is standard for homes)
Power in kVA = 0.5/0.8 = 0.625

An inverter of a standard rating of 1 kVA is required to carry the loads above.

Battery demand for 12 h of working
The battery support duration in an inverter scheme completely depends upon the battery capacity (in Ah, Ampere-hour) as well as the number of batteries required as per the load requirement.

Battery backup time and the number of batteries are calculated as:
Let T_b = battery backup time in hours
C = battery capacity in AH
V_b = battery voltage in volts
N_b = number of batteries in series or parallel as the case may be.
P_L = connected load in watts (W)
Now

$$T_b = (C \times V_b \times N_b)/P_L \tag{7}$$

$T_b = (200 \times 12 \times \text{N})/500 = 12\,\text{h}$
$N_b = 2.5 \approx 3\,\text{Numbers}$

This shows the minimum battery requirement for more than 12 h of backup per day that is a secondary option for the proposed system reinforced in the distributed network. Storage energy for the control and management of street light or traffic light or CCTV is based on the optimal operation scheduling of the system. Normally, the power consumption of CCTV and LED traffic light is much less ($\approx 30\%$ less) as compared to the street light; hence IOT-based smart monitoring system will provide greater benefit in the direction of energy management.

3. IOT devices for automatic light control

Improper management of street lights leads to more power consumption and inefficient as it is working for a whole night, even though not necessarily required in many of the time during the absence of vehicles or pedestrians [13]. Smart street light basically focuses on energy-saving scheme through IOT-based automatic light intensity control technology during night hour. This will make possible using more number of street lights than normal case or the energy can be diverted for BES to utilize in external circuit applications for monitoring traffic light or CCTV.

3.1 Circuit connections

Connection of the proposed system for automatic street light control is depicted in Fig. 3 which consists of ESP8266 Wi-Fi module with 1 MB of build in flash, LDR, ultrasonic senor, and Arduino IDE software in addition to that dimmer circuit used to control 16 dimming levels controlled using controller, switches, and PLC logics. Preliminary experiment done through breadboard, ultrasonic sensor, and LDR both connected to Wi-Fi module, where GND and 3V3 pins of the ESP8266 Wi-Fi module, respectively, connected to the ground and VCC pin of ultrasonic sensor. ESP8266 works by power-saving algorithm to operate with 5 subsequent modes such as active, wake-up mode, sleep mode, deep sleep mode, and OFF [14]. "Echo" and "Trig" pins of ultrasonic sensor are connected to the "GIPO5" and "GIPO6" pins of the Wi-Fi module, respectively,

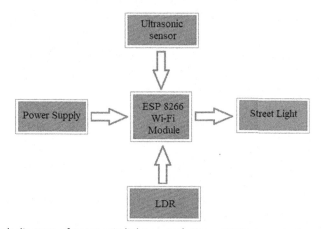

Fig. 3 Block diagram of automatic light control. *No permission required.*

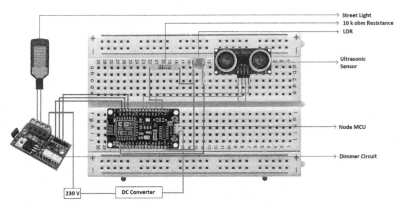

Fig. 4 Connection diagram. *No permission required.*

whereas both wires of LDR are connected to the A0 pin (ADC pin) and GND through 10 KΩ resister of the Wi-Fi module, respectively. Dimmer module is connected to the o/p of GPIO pins 0, 2, 13, and 4 of Wi-Fi module to work on different dimming levels smoothly with accurate firing angle control. Circuit connection was made as per the depicted diagram in Fig. 4.

3.2 Working

During day time, LDR is working with maximum resistance, but during night time, resistance decreases to a minimum level which allows the LED light to glow with its 60% efficiency whose resistance value is programmed with Arduino IDE. Ultrasonic sensor determines the movement of object from certain distance; it activates the switching circuit when any object is coming within the range of sensor, and hence, if there is any presence of object, it allows the bulb to glow with its 100% efficiency. SSR 230V 4A Dimmer circuit ensures the smoothness of power supply converted through DC converter. When there is no object found within its range, it continues to glow with 60% efficiency during night hour and off during day hour that saves nearly 40% of the electricity used than in normal case.

4. Conclusion and result analysis

In the proposed system, idea behind the smart town, its significance, and obtainable result in metropolitan through IOT completion has been described. Energy efficient equipment and its implementation in proposed system are also mentioned. The main purpose of the system is utilization of waste water to produce WtE through Pico-hydro and controlling of street

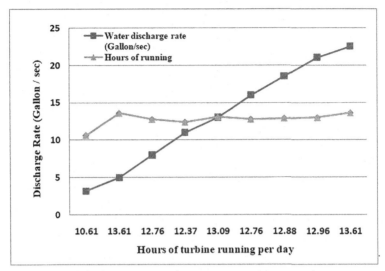

Fig. 5 Waste water discharge rate per hour. *No permission required.*

light by using IOT-based sensors. Estimation of water available is the main constraint for continuous power generation for which a precise analysis was done by considering few residential houses and their daily water usage that was randomly monitored from their overhead tank. A minimum limit of water discharge is considered for the better analysis of result as shown in Fig. 5. It is shown that in some cases the working duration of Pico-hydro turbine is exceeding more than 12 h, which signifies that availability of abundant water for power generation is more than desired quantity.

The number of street lights connected is based on the power generation and optimization methods. Optimal utilization is made through the IOT-based power monitoring scheme by which the energy usage capacity can be enhanced up to 60% as compared to the direct use of street lights. Hence with minimum energy, street lights can be operated by smart monitoring system. Fig. 6 shows comparative analysis between the energy usage during normal condition and through IOT devices which figures out less electricity consumption and leads to better revenue generation with reduced payback period. WtE generation concept is also reducing the environmental pollution through proper utilization of treated water for watering in garden or plant, and recycling of water can be done through partial utilization of generated power.

It reduces the amount of energy wasted associated with the manual streetlight switching. By the use of LDR, it creates an effective and

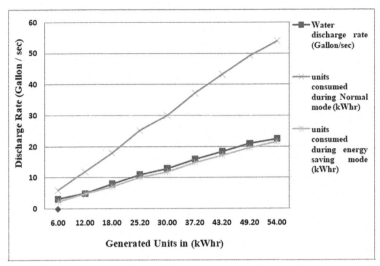

Fig. 6 Energy generation in different conditions. *No permission required.*

intelligent street lights monitoring system. It has the potential to cut energy usage and maintenance requirements. It can be used in both urban and rural settings. The new method is expandable and completely customizable to the user's requirements. When needed, it generates a safe place with higher intensity light. The system's purpose is to minimize ongoing costs while also increasing the system's longevity. The system's upfront costs and upkeep are among its drawbacks.

References

[1] S. Pellicer, G. Santa, A.L. Bleda, R. Maestre, A.J. Jara, A.G. Skarmeta, A global perspective of smart cities: a survey, in: Proceedings—7th International Conference on Innovative Mobile and Internet Services in Ubiquitous Computing, IMIS 2013, 2013, pp. 439–444, https://doi.org/10.1109/IMIS.2013.79.

[2] J. Jin, J. Gubbi, S. Marusic, M. Palaniswami, An information framework for creating a smart city through internet of things, IEEE Internet Things J. 1 (2) (2014) 112–121,- https://doi.org/10.1109/JIOT.2013.2296516.

[3] Y. Fujii, N. Yoshiura, A. Takita, N. Ohta, Smart street light system with energy saving function based on the sensor network, in: e-Energy 2013—Proceedings of the 4th ACM International Conference on Future Energy Systems, 2013, pp. 271–272, https://doi.org/10.1145/2487166.2487202.

[4] D. Evans, The Internet of Things: How the next evolution of the Internet is Changing everything, CISCO, 2011.

[5] A.B. Chhetri, G.R. Pokharel, M.R. Islam, Sustainability of micro-hydrosystems—a case study, Energy Environ. 20 (4) (2009) 567–585, https://doi.org/10.1260/095830509788707356.

[6] J. Titus, B. Ayalur, Design and fabrication of in-line turbine for pico hydro energy recovery in treated sewage water distribution line, Energy Procedia 156 (2019) 133–138, https://doi.org/10.1016/j.egypro.2018.11.117.

[7] D. Giusto, A. Iera, G. Morabito, L. Atzori, The Internet of Things, 17, Issue 6, Springer, New York, 2010, https://doi.org/10.1007/978-1-4419-1674-7.

[8] P.M. Santos, J.G.P. Rodrigues, S.B. Cruz, T. Lourenço, P.M. D'Orey, Y. Luis, C. Rocha, S. Sousa, S. Crisóstomo, C. Queirós, S. Sargento, A. Aguiar, J. Barros, PortoLivingLab: an IoT-based sensing platform for smart cities, IEEE Internet Things J. 5 (2) (2018) 523–532, https://doi.org/10.1109/JIOT.2018.2791522.

[9] M.S.S. Muthanna, et al., Development of intelligent street lighting services model based on LoRa technology, Proceedings of the 2018 IEEE Conference of Russian Young Researchers in Electrical and Electronic Engineering (EIConRus), IEEE, 2018.

[10] S. Priyanka, Control of solar LED street lighting system based on climatic conditions and object movements, J. Inf. Knowl. Res. Elect. Eng. 3 (2) (2014) 480–486.

[11] R. Kodali, S. Yerroju, Energy efficient smart street light, Proceedings of the 2017 3rd International Conference on Applied and Theoretical Computing and Communication Technology (iCATccT), IEEE, 2017.

[12] Homewateruse. (2019). http://www.phila.gov › Documents › Homewateruse_IG5.

[13] Deepak, Smart Street Lights by Srivatsa, Preethi, Parinitha, & Sumana, Kumar BNM Institute of Technology, 2013.

[14] Kolbans Book on ESP8266, Kolbans Book on ESP8266, an introductory book on ESP8266, 2015.

PART THREE

Smart education

Reigniting the power of artificial intelligence in education sector for the educators and students competence

Abdul Razak[a], M. Pandya Nayak[b], Geetha Manoharan[a],
Subhashini Durai[c], Gunaseelan Alex Rajesh[d], Col B.S. Rao[a], and
Sunitha Purushottam Ashtikar[a]

[a]School of Business, SR University, Warangal, Telangana, India
[b]Department of Commerce, PG Centre, Palamuru University, Mahabubnagar, Telangana, India
[c]GRD Institute of Management, Dr. G.R. Damodaran College of Science, Coimbatore, Tamil Nadu, India
[d]Sri Venkateswara Institute of Information Technology and Management, Coimbatore, Tamil Nadu, India

1. Introduction

Through an evaluation of policies and private sector activities, the present study investigates the relevance of artificial intelligence (AI) and education in India. While the media and officials portray India's AI development as a well-coordinated national plan and a developing geopolitical war for global dominance, this is not the truth. This investigation will result in a more sophisticated internal complexity that includes a variety of regional networks and worldwide commercial activity. The present research emphasizes the need for including educational institutions as key stakeholders in national and regional AI development initiatives, with an emphasis on educating local experts. The bulk of private-sector educational institutions are putting out significant effort in this area.

Artificial intelligence (AI), machine learning (ML), and deep learning (DL) are revolutionizing education, approaches, tools, systems, and institutions as educators, psychologists, and parents debate how much screen time is appropriate for children. This is producing changes in the education business and forcing us to think about the future of education. The artificial intelligence market in the education sector research predicts that artificial intelligence in education will increase in all industries, including education. Machine learning (ML) and artificial intelligence (AI) are crucial drivers of progress and innovation. Over 47% of learning management products will include AI capabilities in the next three years, according to eLearning Industry. Despite the fact that most experts believe that teachers' importance cannot be overstated, the teaching profession and educational best practices will undergo significant changes.

The EdTech industry has been slow to adopt AI-powered solutions, despite the fact that they have been available for some time. In contrast, the pandemic drastically altered the environment, compelling instructors to rely on technology to provide virtual training. Technology, according to 86% of teachers, should now be a major component of classroom instruction. Artificial intelligence (AI) has the potential to improve the teaching and learning process, assisting educators evolving in ways that benefit both instructors and students.

Artificial intelligence (AI) is the practice of simulating human cognitive processes with technology and tools, particularly computer systems. These processes consist of learning, or the acquisition of information and the rules for applying it, reasoning, or the application of rules to approximate conclusions, and self-adjustment. A wide range of educational stakeholders will benefit from the research. It will contribute to the expanding body of knowledge, theory, and empirical data regarding the effects of artificial intelligence on education. It will aid researchers, professionals, and policymakers, such as administrators, managers, and leaders of academic institutions and the education sector, by developing fact-based decision-making, management, and leadership processes in the sector.

2. Significance of the study

The act of using computers and machines to complete a task by replicating human perception, decision-making, and other processes at the most fundamental level is known as artificial intelligence (AI). To put it another way, AI is the use of computers to perform complex pattern matching and learning tasks. Human judgement will never be completely replaced by AI, but it is getting there. Teachers can now automate the scoring of nearly any number of multiple-choice and fill-in-the-blank tests, and automatic grading of student work is not far behind. Academic paper technology is still in its infancy and is far from perfect, but it has the potential to improve over the next few years, allowing instructors to devote more time to in-class activities and student participation rather than grading. The current study emphasizes artificial intelligence's importance in the education industry as a pressing need.

3. Need of the study

In this modern era, machine-based AI is already extensively used in education. Natural language processing is used by a number of testing businesses to evaluate essays. MOOCs, such as Edx and Coursera, which allow for limitless participation through the internet, have also employed AI grading to score papers inside their courses. Natural language processing is increasingly being used by schools and universities to evaluate the essay component of their yearly assessment. It helps in saving money on exams by using this strategy. Several academics are experimenting with ways that integrate AI, machine learning, and natural language processing to produce new, standardized exam items automatically based on a large body of information and evidence. The technologies make use of artificial intelligence and agility, and the curriculum and material have been designed to meet the demands of students, resulting in increased absorption and perseverance, as well as improved comprehension and understanding.

4. Objectives of the study

The study focuses on artificial intelligence in the education sector and how AI is used to improve education. The study's other goals are as follows:
➤ To study how the artificial intelligence can be adapted with the students and educators needs.
➤ To examine how artificial intelligence-driven courses can give students and educators helpful results.

5. Scope of the study

The study's scope is limited to artificial intelligence in the education sector and its usage in education by both the teachers as well as students. As a result of the epidemic, most educational institutions began to offer online courses, paving the way for AI to enter the education industry. This study focuses on AI-driven education courses. Artificial intelligence (AI) began with computers and related technology and has since expanded to encompass all aspects of life. The research is confined to how teachers use artificial intelligence to educate pupils in the educational sector. The study's focus is confined to educational institutions, teachers, and students' use of new and emerging technologies in the education sector.

6. Review of literature

According to Chen, Chen, and Lin [1], Artificial intelligence (AI) is quickly becoming accepted and used in a variety of ways in education, particularly by educational institutions. Instructors have been able to complete administrative tasks such as evaluating and grading students' assignments more quickly and efficiently, as well as improve the quality of their teaching methods, thanks to these platforms.

AI has always made a significant contribution in the field of education, according to Malik, Tayal, and Vij [2]. AI has always aided both instructors and students, from robotic teaching to the development of an automated scoring system for answer papers. All natural language processing-enabled intelligent tutor systems need artificial intelligence to function. These techniques improve self-reflection, deep questioning, conflict resolution, creative inquiry, and decision-making ability.

Pedro, Subosa, Rivas, and Valverde [3] looked at how artificial intelligence (AI) is applied in education throughout the globe, especially in underdeveloped countries. It also sows the seeds of discussion and thinking during mobile learning week and beyond, as well as many other approaches to achieve SDG 4, which is centered on education. Studying how governments and educational institutions are rethinking and reforming educational processes to better prepare students for the growing presence of AI in many aspects of human life is the focus of the book "Preparing Learners to Prosper in an AI-Saturated Future."

Zawacki-Richter et al. [4] stated artificial intelligence in education signifies a turning point in educational technology, although it is yet unknown in teaching and learning. The purpose is to figure out how to overcome the obstacles to AI research in higher education, and the bulk of the data shows that AI education encompasses a wide variety of fields, from computer science to STEM. The four domains of AI education addressed in this research include profiling and predictions, measurement and evaluation, optimum efficiency and customization, and integrated learning systems. Finally, it recommends that the lack of critical thinking about concerns and risks be redirected and pondered upon.

Baker [5] claimed that the near future of artificial intelligence and educational studies will be constructed on three educational process applications: models as scientific instruments, models as features of instructional artefacts, and models as teaching product design grounds. This phrase refers to all current attempts to explore learning settings as well as several theories and models in individual cognition. In the second function, computer-based learning systems have been pushed for inclusion in courses where students "open up" technology in education and "open up" educational technologies to performers, with those performers "opening up" to technology. In their third purpose, they also provided design methodologies and system components for a variety of tools available to students. Finally, each of the three responsibilities should be involved in some manner with each case.

Chassignol et al. [6] stated digital technology has become an indispensable element of our everyday life. They influence how we see the world, how we communicate knowledge and information, and how we behave. This research focused on four categories: personalized educational content, innovative teaching strategies, technology-assisted assessment, and the student-lecturer relationship.

Muzammul [7] stated that as per artificial intelligence, re-engineering presents educational institutions with the use of cutting-edge technology may be highly advantageous in terms of quality systems. Artificial intelligence is employed in this research to achieve re-engineering based on two theories: multiface recognition (MFR) and facial expression recognition (FER). Both hypotheses are supported by intelligent systems like principal component analysis, discrete wavelet transform (DWT), and k-nearest neighbor (KNN). The constructed system can distinguish expressions such as happiness, disgust, fear, fury, and confusion. The expert system and knowledge base are two applications for ensuring educational quality.

According to Jordan and Mitchell [8], AI and machine learning (ML) create systems that automatically learn from their experiences, and it is one of the fastest-growing technical fields, straddling the lines between computer science and statistics, as well as artificial intelligence and data science. Machine learning is being propelled forward by recent advances in innovative learning algorithms and theory, as well as the ongoing proliferation of online data and low-cost computing. Machine learning methods are used to make decisions in a variety of industries, including health care, manufacturing, education, financial modeling, law enforcement, and marketing.

Carleo et al. [9] stated that in recent years, machine learning has produced a wide variety of algorithms and modelling tools that are used for a variety of data processing tasks and serve as a bridge between machine learning and physical sciences. It entails conceptual improvements driven by physical discoveries, machine learning applications to a wide range of physics domains, and cross-pollination. This paper discusses statistical physics methods and concepts, as well as applications of machine learning methodologies in particle physics and cosmology, quantum many-body physics, quantum computing, and chemical and material physics. The focus is on ML, which is aimed at progressing toward new computer architectures and discusses all existing triumphs as well as domain-specific approaches and difficulties.

Mahesh [10] says that machine learning is the study of algorithms and statistical models in which computer systems are used to do a task without being explicitly programmed, and learning algorithms in applications are used in everyday life. Data mining, image processing, predictive analytics, and a variety of other applications use algorithms. Machine learning has the primary advantage of automatically processing data.

Machine learning, according to El Naqa and Murphy [11], is a rapidly developing field of computer algorithms that aims to replicate human intelligence by learning from their surroundings, and it is sometimes referred to as the workhorse of the so-called big data era. In this study, radiotherapy is defined as a complex set of activities that go from the consultation phase to the treatment phase and include multiple phases of complex interactions between humans and machines and decision-making. The ability of machine learning algorithms to learn from their current environment and generalize to previously encountered tasks allows for increased radiation safety and efficiency, resulting in better outcomes.

7. Research methodology

The research methodology primarily focuses on the data sources, data gathering methods, research methodologies used, and research instruments and techniques employed for data processing and interpretation.

7.1 Types of data

The data are acquired from two sources: primary and secondary sources. The data are mostly gathered from secondary sources, which include papers, journals, books, and websites, among others. The experts responses are the primary source of data, and their views are taken into account.

7.2 Research method

This is a research project aimed at reviving artificial intelligence's potential in the education sector in order to improve efficiency. Case study methodology is chosen.

7.3 Limitations of the study

The research is confined to artificial intelligence in the field of education. This research aims to discover how artificial intelligence (AI) may help educators and students work more efficiently. Because artificial intelligence is a new phrase that is being used for the first time, the research will only look at recent advancements.

8. Theoretical framework

One of the most important purposes of AI in education is to provide tailored mastering guidance or assistance to character college students based on their mastering popularity, potential, or personal characteristics. Allowing learning structures to serve as a shrewd show by incorporating skilled teachers' knowledge and intelligence into the system's decision-making process is a critical problem from the perspective of precision schooling, which emphasizes the need to provide prevention and intervention practices to individual beginners by studying their learning status or behaviors. The subject of intelligent tutoring structures (ITSs) was highlighted by instructional period and laptop technology researchers in the early 1980s. Designing intelligent tutoring systems and adaptive learning systems faces challenges that go beyond computer programming skills. We use human tutors' knowledge and experience as well as the best available evidence to make the best possible decisions and choices for each individual learner. The AIED's strong reliance on technology and cross-disciplinary nature creates these challenges. If researchers do not understand the roles of AI in education and how AI technologies work, they may be unable to properly deploy AIED applications and activities, as well as raise and investigate important AIED research concerns. The ideas presented here are meant to spark discussion about what would be viable pathways for future AIED study instructions, and I understand that some of you may have already done research along these lines, so these concepts may not be new to anyone.

Zhao and Liu [12] compared the various AI training systems throughout the arena. They talk about Smart Sparrow, which is an online adaptive training platform in Australia. They also include Desire2Learn (D2L), a cloud school learning management platform located in the United States, as well as Jill Watson, an online AI program based on the IBM Watson supercomputer. Betty's Brain, an AI, is also mentioned by Zhao and Liu.

Preliminary research revealed a narrative and framework for evaluating AI that focused the investigation on the utility of AI in management, guidance, and learning.

- A qualitative research strategy was adopted, utilizing literature review as a study design and technique, which successfully supported the examiner's belief.
- The first section of this disc examines how artificial intelligence (AI) may be utilized to enhance mastering outcomes. It gives instances of how artificial intelligence (AI) might help schooling systems utilize data to promote educational fairness and quality in developing countries.
- The second section, "Preparing Newcomers to thrive in an AI-Saturated Future," examines how governments and educational institutions are rethinking and rewriting educational programs to prepare inexperienced people for AI's expanding involvement in all aspects of human life.

However, it is unrealistic to believe that to build fully functional educational robots or extend a smart lecture hall in the next few years, but we will be able to chart a course in that direction. Engaging in this line of action might result in tremendous development for the sphere and new partnerships with sectors such as robotics and electrical engineering. Future research topics could include developing structures that allow for socialization (as an extension of the already-existing computer supported collaborative learning); discovering different modalities for integrations between newbies, instructors, and assistive technologies; and looking into the software of the Internet of Things in education. Robots will be able to learn much of what we know about how to model knowledge and what the learner knows, as well as how to provide feedback and teach. As a result, additional techniques of perceiving the learner's emotional realm and connecting with them in various ways might be introduced. This might create new challenging conditions for AIED, which could be incredibly interesting and allow us to expand our models and techniques.

9. Analysis of artificial intelligence in education sector

(a) **Basic educational tasks, such as grading, may be automated using artificial intelligence:** Even with the help of teaching associates, grading large lecture courses' assignments and exams may be a time-consuming task. A lot of time is wasted on grading even in the lower grades, which may be used for teaching or preparing courses or working on professional development. Human grading may never be totally replaced by artificial intelligence, but it is getting close. Teacher evaluations could soon be as simple as multiple choice and fill-in-the-blank exams, if not easier. There is a long way to go before essay-grading software is up to par, but it will enable professors and instructors to spend more time on in-class activities and student involvement than on grading now.

(b) **Artificial intelligence can make trial-and-error learning easier:** However, many students are afraid of failing or just not knowing the answer, despite the importance of experimentation in learning. Putting yourself on the spot in front of your classmates or professors might be uncomfortable for some individuals. Using a computer system designed to help children learn is a far less daunting way to deal with trial and error. If AI educators can provide ideas for improvement, they may give students with a safe atmosphere in which to experiment and learn. For this form of learning, AI is the optimal framework, since AI systems often learn through trial and error.

(c) **Artificial intelligence may discover opportunities for improvement when it comes to courses, programs, and instructional aspects:** It is possible that teachers are not aware of the fact that some sections of their lectures and instructional materials leave students perplexed. This problem can be solved with the help of artificial intelligence. Open online course providers, such as MOOCs, already accomplish this. When a large number of students submit wrong answers to a homework assignment, the system alerts the teacher and delivers personalized messages to future students with instructions on how to remedy the issue.

(d) **Students may benefit from the help of artificial intelligence instructors:** Human instructors, as opposed to those who use artificial intelligence (AI), have a number of advantages. Tutors who only exist in binary may become increasingly common in the near future. Apps that use artificial intelligence to help students with arithmetic, writing, and other subjects are already available. Real-world instructors must still support the development of higher-order thinking and creativity, even if these programs can teach the fundamentals. AI educators may be able to do these duties in the future, although this has not been shown yet. Advances in teaching methods may not be a pipe dream because of the rapid advancement in technology over the last several decades.

(e) **Students will be better served as a result of AI-enabled data being integrated into educational systems:** Colleges are rethinking how they connect with prospective and current students because of sophisticated data collecting and computer systems. Recruiting new students and assisting them in their course selection are just two of the many ways computers are making college more student-centered.

In today's higher education, data mining tools are already prevalent, but artificial intelligence has the potential to revolutionize the field. AI-guided training is already being tested in several institutions to assist students in the transition from high school to university. We may one day see a college admission process similar to Amazon or Netflix, with a system that suggests the best colleges and programs to students based on their interests.

(f) **Artificial intelligence (AI) is changing how we find, discover, and engage data:** On a daily basis, many of us do not even notice the algorithms that impact the information we see and discover. For example, Flipkart and Snapdeal recommend products based on previous purchases, Google Assistant and Siri respond to your queries, and Internet searches based on your location. In our personal and professional lives, intelligent systems like these play an essential role in how we access and utilize information, and they may soon revolutionize the way we access and use information in schools and academic institutions. There has already been a dramatic shift in the way people interact with information over the last several decades, and newer, more integrated technology might have a profound influence on how students investigate and verify information in the future.

(g) **Students' needs and demands can be met by customizing or tailoring educational software:** From basic through postgraduate levels, personalized learning will be more important as artificial intelligence (AI) becomes more prevalent. Adaptive learning programs, games, and software are all on the rise, so some of this is already taking place in the real world. Systems like this may be customized to meet the needs of individual students, focusing more on certain subjects, reiterating concepts that students do not understand, and allowing them to study at their own pace.

(h) **Artificial intelligence has the potential to alter instructors' roles:** Technology, such as intelligent computer systems, may change the nature and scope of teachers' roles in education. Grading, helping kids learn better, and even substituting for in-person tutoring are all things that AI can accomplish, as we have previously proven. AI, on the other hand, might be applied in a wide range of educational contexts. AI systems might be taught to provide expertise in relatively simple course topics, allowing students to ask questions and gain information, or perhaps assuming the place of teachers. Although in most cases, AI will replace the teacher's role with that of a facilitator; this is not always the case.

Instructors will use AI-assisted education to improve, aid students in need, and give them with human engagement and hands-on experiences. Especially in online or flipped classrooms, many of these shifts are already being driven by technology in the classroom.

(i) **Artificial intelligence has the potential to alter how students learn, who teaches them, and how they acquire basic skills:** Decades pass before we see major changes, but artificial intelligence (AI) possess the skill to fundamentally revolutionize almost in every aspect of education in the near future. The use of AI systems, software, and help means that students may study from anywhere in the world at any time, and in certain circumstances, AI may even replace professors in some cases (for better or worse). Students are presently receiving assistance from educational AI systems, but as these programs grow and their authors acquire more expertise, they will be able to provide students a wider variety of services in the future. Education may look substantially different in a few decades from now if current trends continue.

(j) **Students and instructors can benefit from artificial intelligence-driven tools that provide constructive and valuable feedback:** There are a number of ways in which artificial intelligence (AI) might help educators and students design courses that are personalized to their individual needs. In certain universities, especially those that offer online courses, artificial intelligence technologies are being used to monitor student progress and alert professors when a student is not meeting expectations.

Students may benefit from these AI systems, and teachers may be able to see where they can improve teaching for students who are struggling. These universities' AI programs, on the other hand, are not just there to help students get through certain courses. Some people want to develop methods that make students to choose their majors based on their skills and weaknesses. When choosing a college major in the future, students are not obligated to follow the advice, but it could open up a whole new world of possibilities.

10. Conclusion

Incorporating artificial intelligence in education (AIED) into smart classrooms could generate real-time data streams from multiple sensors in the study room and focus on the newbie, which could necessitate a lot of educational data mining resulting in updated models of how newbies energize in the classroom's wider environment, rather than just with the educational applications we have developed so far. At the artificial intelligence in education (AIED), or a related conference, it may be possible to keep a session on educational robots and smart classrooms in place. You now have complete control on where the artificial intelligence in education (AIED) goes next.

As a result, institutions should re-examine their roles and pedagogical methods, as well as their relationship to AI solutions and their owners, in the near future. In addition, institutions of higher education are well aware of the wide range of possibilities and challenges that AI may bring to the table when it comes to teaching and learning. There are new opportunities for training for everyone, as well as lifelong learning and the preservation of center values and higher education's mission, thanks to these responses. As we consider the moral implications of today's management of AI advances, we must keep in mind that there is the potential for the monopoly of a few corporations to diminish the variety of human knowledge and viewpoints. Further research on the new responsibilities of instructors in new learning paths for higher-degree students, acquired with an enhancement of graduate qualities, with a focus on creativity and invention, the set of skills and abilities that will never be recreated by computers, are also critical. Since artificial intelligence (AI) is on the rise, it is impossible to overlook an important discussion about how institutions will choose students to learn about AI's potential function as a teacher and guide in higher education. As a result of the rapid rate of technological change and employment displacement (supply), teaching in higher education requires an overhaul of instructors' positions and instructional methods.

Technology-based responses, such as "getting to know management structures" or "IT solutions to stumble on plagiarism," have already improved the issue of who sets the coaching and learning schedules. A dismal future is made more likely by the development of tech lords and the quasimonopoly of a small number of tech firms. Due to this collection of hazards, institutions must consider a sustainable future while dealing with these issues. Furthermore, AI software built entirely on intricate algorithms developed by programmers and capable of transmitting their own biases or agendas into operating systems will replace many current units of responsibilities at the center of training exercises in improved training. Universities' ability to preserve civilization, foster knowledge, and grow knowledge and competence depends on their ability to conduct continual critiques and investigations into offered solutions.

References

[1] L. Chen, P. Chen, Z. Lin, Artificial intelligence in education: A review, IEEE Access 8 (2020) 75264–75278.

[2] G. Malik, D.K. Tayal, S. Vij, An analysis of the role of artificial intelligence in education and teaching, in: Recent Findings in Intelligent Computing Techniques, Springer, Singapore, 2019, pp. 407–417.

[3] F. Pedro, M. Subosa, A. Rivas, P. Valverde, Artificial intelligence in education: Challenges and opportunities for sustainable development, United Nations Educational, Scientific and Cultural Organization, France, 2019, pp. 1–46.

[4] O. Zawacki-Richter, V.I. Marín, M. Bond, F. Gouverneur, Systematic review of research on artificial intelligence applications in higher education—Where are the educators? Int. J. Educ. Technol. High. Educ. 16 (1) (2019) 1–27.

[5] M.J. Baker, The roles of models in Artificial Intelligence and Education research prospective view, J. Artif. Intell. Educat. 11 (2000) 122–143.

[6] M. Chassignol, A. Khoroshavin, A. Klimova, A. Bilyatdinova, Artificial Intelligence trends in education: a narrative overview, Proc. Comput. Sci. 136 (2018) 16–24.

[7] M. Muzammul, Education System re-engineering with AI (artificial intelligence) for Quality Improvements with proposed model, ADCAIJ: Advances in Distributed Computing and Artificial Intelligence Journal, second ed., 8, Ediciones Universidad de Salamanca, Spain, 2019, pp. 51–60.

[8] M.I. Jordan, T.M. Mitchell, Machine learning: Trends, perspectives, and prospects, Science 349 (6245) (2015) 255–260.

[9] G. Carleo, I. Cirac, K. Cranmer, L. Daudet, M. Schuld, N. Tishby, et al., Machine learning and the physical sciences, Rev. Mod. Phys. 91 (4) (2019), 045002.

[10] B. Mahesh, Machine learning algorithms—a review, Int. J. Sci. Res. (IJSR) 9 (2020) 381–386.

[11] I. El Naqa, M.J. Murphy, What is machine learning? in: Machine Learning in Radiation Oncology, Springer, Cham, 2015, pp. 3–11.

[12] Y. Zhao, G. Liu, How do teachers face educational changes in artificial intelligence era, in: 2018 International Workshop on Education Reform and Social Sciences (ERSS 2018), Atlantis Press, 2019, pp. 47–50., January.

CHAPTER EIGHT

A study of postgraduate students' perceptions of key components in ICCC to be used in artificial intelligence-based smart cities

Geetha Manoharan[a], Subhashini Durai[b], Gunaseelan Alex Rajesh[c], Abdul Razak[a], Col B.S. Rao[a], and Sunitha Purushottam Ashtikar[a]

[a]School of Business, SR University, Warangal, Telangana, India
[b]GRD Institute of Management, Dr. G.R. Damodaran College of Science, Coimbatore, Tamil Nadu, India
[c]Sri Venkateswara Institute of Information Technology and Management, Coimbatore, Tamil Nadu, India

1. Introduction

When it comes to growth and expansion in any industry, efficiency is critical. A Command Control Center will improve the lives of GCC citizens by providing them with better understanding and information. Intelligent cities are designed to include the installation of a number of smart components in conjunction with the establishment of an Integrated Command Control Center, according to the Smart Cities Initiative. In order to manage the city's safety and surveillance from a Police or Traffic Police perspective, as well as to host Smart Solutions for the Municipal Corporation, an Integrated Command and Control Center (ICCC) will be established. An Integrated Command Control Center (ICCC) will be established to manage the city's safety and surveillance while also hosting Smart Solutions for the Municipal Corporation. Fig. 1 shows the major or important highlights of the operations of ICCC.

Other than the abovementioned highlights, the integrated smart city framework includes the following facilities also (Fig. 2).

2. Integration of Command and Control Center

In order to function as the brains of city operations, the Integrated Command and Control Centers must be capable of resolving problems and managing disasters. The use of sensors and edge devices will allow for

Artificial Intelligence and Machine Learning in Smart City Planning Copyright © 2023 Elsevier Inc.
https://doi.org/10.1016/B978-0-323-99503-0.00003-X

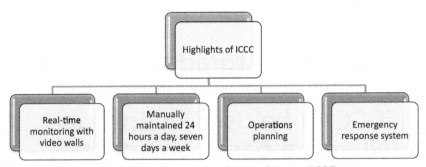

Fig. 1 Highlights of Integrated Command and Control Center (ICCC).

Fig. 2 Facilities included in integrated smart city framework.

the collection and generation of real-time data in the areas of water, waste management, energy, mobility, the built environment, education, healthcare, and safety. As a decision support system (DSS) for city administration, the ICCC platform will serve as a decision support system (DSS) through its various layers and components. This will allow the city administration to respond to real-time events by consuming data feeds from a diverse range of data sources and processing information extracted from the data sets.

The purposes of utilization of ICCC

- Increasing civic officials' situational awareness across urban functions by providing insights gleaned from data collected through the deployment of sensors throughout the city.
- Implementing standard response protocols for recurring events, issues, and emergency scenarios at the municipal level by institutionalizing standard processes for recurring events, issues, and emergency scenarios.
- Collaboration between multiple departments within and outside of urban local governments and government agencies is essential for success.
- When it comes to routine operations and during times of crisis, data-driven decision-making is becoming more institutionalized at all levels of city functionaries from operators to city administrators.
- Engaging support staff on the ground to address civic issues and citizen complaints.

3. Need for ICCC assessment

These facilities have received significant investment; however, they have yet to be hardwired into the urban local body's day-to-day urban management functions. Without a reference model for best operational practices and a lack of guidance on service integration, there appears to be widespread ambiguity among completed ICCC projects regarding their current status and intended outcome. The evaluation model provides a simple form of benchmarking the ICCC ecosystem in smart cities across the country. These benchmarks would be running in parallel with the livability standard and categories of indicators and will further cover the essential aspects of an ICCC that is optimized in terms of the utilization of the governance, and technology framework to help make the city more livable and developed.

Additionally, the assessment will assist smart cities in determining the maturity of existing ICCCs investments and identifying areas for improvement. It will also act a launch pad for the cities which are yet to commence their city operations and system integrations. The assessment will allow cities to understand the impact of ICCC on the following parameters related to livability indicators, as well as identify how urban operations have improved over time, in addition to the traditional way of managing operations.

4. MoHUA livability index at smart cities

The Government of India, with the assistance of various State and Local bodies, is implementing a number of flagship urban development programs. In order of making Indian cities more livable, MoHUA had launched one such initiative of ranking cities using a livability index which measures parameters across urban domains. All cities are striving to achieve a good score on this index, which gauges the quality of life for the citizens through improvements across multiple dimensions like housing, transportation, utilities, mobility, ICT, health, education, and economy. The Integrated Command and Control Center, if operationalized and managed successfully, can play a pivotal role in improving the livability of a city by ensuring efficient service delivery and quicker response to emergencies or crisis situations. It will in turn assist in improving the benchmark scores of cities on livability index by enhancing the monitoring of city services, data collection and analysis, based on which the ULBs can make informed decisions leading to better quality of life for the citizens. Hence, it is imperative that the functions and operations of the ICCCs are aligned with key performance indicators as defined under livability indicators that capture the extent and quality of infrastructure, service delivery (water supply, SWM services, healthcare, and e-governance), and emergency response services.

5. Current process of implementation of ICCC

The true value of an Integrated Command and Control Center is realized through the optimization of city operations and the ability to make informed decisions. Additionally, there is no uniformity in the implementation models that are adopted around the country. In many cities, departments or organizations have decided to set up their own Command and Control Center to deliver one particular function (for instance using Command and Control Center for the police to look after safety and security or to manage water related function), whereas on the other hand, selected cities have aggregated entire city operations under an Integrated Command and Control Center. While the city may choose the hub and spoke model or integrated model based on the way ULB functions are organized, it is important that the Integrated Command and Control Center is used for viewing, correlating, commanding, and controlling city operation including day-to-day scenarios and use case and exception management.

6. Architecture of ICCC

It consumes real-time data from sensors devices and data sources as well as static and real-time data feeds from various applications, systems, and databases, among other things. Examples of sensors devices include air and water quality sensors for street light management, metering devices, telemetric and location-based devices, proximity sensors, surveillance and safety cameras, disaster detection sensors, level sensors for solid waste management, and so on. Data collected by sensors and processed by various components of an ICCC can be used to generate information, which can then be used to generate alerts. Alternatively, an ICCC can connect to COTS and be spoke applications in order to generate alerts generated by the integrated COTS and bespoke application/systems. In order to obtain information, this layer enables other ICCC components to aggregate, consume, and process data in order to obtain information from the data.

The data aggregation and analytics layer is in charge of extracting information and intelligence from data collected by the data acquisition layer from a variety of data sources and storing it in a centralized location. When it comes to data aggregation and analysis, it is a collection of components that make it possible to extract and transform information from a wide range of systems and data sources in a variety of formats. Health records are collected by the Integrated Hospital Management System, traffic data are collected by the Adaptive Traffic Management System, and ambulances are tracked by the Vehicle Tracking System, which collects data in a variety of formats. In addition to processing data, the ICCC data aggregation and analysis layer provides access to data from multiple systems in accordance with the specific requirements of the end users who access the data. Information extraction from a variety of data sets across a domain is accomplished through the use of data analytics components (Fig. 3). By utilizing ICCC components or third-party tools or applications, this intelligence can then be used to handle exceptions and visualize data in a variety of scenarios.

The data acquisition and collection layer provides the ICCC with the ability to extract intelligence from the data acquired and gathered by the previous layer. The business logic application layer is the central application engine of the ICCC platform, and it assists end users in the design and configuration of standard operating procedures, the management of external and

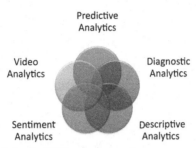

Fig. 3 Role of data analytics in ICCC management.

internal triggers, the implementation of policies, and the handling of complex events. The application layer also assists the ICCC in dealing with real-time events by providing it with intelligence and information from a variety of other systems and sources. As part of its response management, the application layer takes into account various scenarios and their corresponding business logic configurations. It assists with the configuration or automation of operations in a variety of scenarios listed below:

(a) Defining and configuring the event
(b) Defining and configuring the external/inner trigger
(c) Define and configure event response
(d) Define and configure responsibility matrix
(e) Define and configure incidents and change requests
(f) Defining and configuring user access and authorization
(g) Defining and configuring field asset access policy

The ICCC's application layer enables communication with a variety of systems which are as follows.

(i) Configuring events and response for water supply operations:
    a. Configuring alerts and notification using smart metering for water usage or consumption
    b. Configuring events and trigger over data emanating through SCADA system for managing water operations
    c. Configuring response protocol in case of leakage detection and
    d. Configuring response protocol in case of effluent detection
(ii) Event and response management for waste water treatment
    a. Configuring alerts and notification using SCADA for waste water treatment systems for its on-field employees
    b. Configuring alerts and notification using for level detections at treatment plant
    c. Configuring events and trigger for managing energy consumptions of pumping control systems for storm water management

The ICCC utilizes this layer to process events and make real-time decisions in accordance with the protocol configuration.

7. The ICCC's command and control layer is in charge of managing

(a) Keeping in touch with stakeholders
(b) Device management (asset, access and authorization)
(c) Visualization and UI
(d) Complex event handling in real-time situation
(e) Administration of services

To provide city administrators and citizens with actionable information, the command and control layer will house action oriented standard operating procedures, incident response dispatches, and management systems (rules engines, diagnostics systems, control systems, messaging systems, and events handling systems), as well as a reporting or dashboard system. While this layer will be present in the majority of Integrated Command and Control Centers from the start, it will be sufficiently adaptable to accept inputs from various downstream applications and sensors as they are introduced in the city.

8. ICCC maturity assessment framework

ICCC maturity assessment framework assesses the ICCC ecosystem on the following dimensions [1]:
- Functional coverage refers to the city utilities and services such water, waste water, storm water, sanitation, waste management, roads, traffic, and street lights, social infrastructure such as education, healthcare, leisure and recreational facilities, and services such certificate, permissions, and license issuance etc.
- Technological coverage refers to the technological capability, scalability, and security components of the given product.
- Governance essentially refers to people side of the system with emphasis on the governance policies in place to operate the city ICCC efficiently.

Maturity framework assesses the level of maturity of above-mentioned components through a detailed questionnaire to understand how effectively the city has leveraged the capabilities of Integrated Command and Control Center to improve its day-to-day operations, policy, and decision-making. Maturity framework would also help cities in identifying the implementation, technological, and operational gaps under various smart solution projects so that Integrated Command and Control Center could achieve its true potential. Maturity assessment framework would also expose cities to various possibilities of efficient and effective urban governance which impacts day-to-day life of its citizens and officer managing the infrastructure or utilities operations. ICCC effectiveness would be gauged through functional use cases like water supply/solid waste/civic services/emergency management. Objective is to focus on functional aspect of governance through ICCC with objective of bringing all civic bodies operations under one roof to achieve efficiency and effectiveness through standardization, achieve better situational awareness for bringing in optimization, and reduce human touchpoints in providing response and developing capability to manage complex situations in real time through ICCC.

9. Maturity assessment process

ICCC maturity assessment process is a three-step process which is covered in Table 1.

10. Evaluation criteria: ICCC functional capability assessment

The first component of the framework is assessing functional capability of ICCC, i.e., the civic utilities or services being monitored by either the system or the people deployed at the facility. This shall include the services which are supposed to get integrated with the ICCC and covers the primary services given out by the cities like Municipal Utilities and Civic Services [1]. These are primarily civic services provided by the urban local body (ULB) to meet the general public's daily needs. A city ICCC is required by the framework to integrate and monitor these services on-site, as service disruptions and a lack of timely response may result in substandard service delivery. ULBs are in charge of the following essential services.
(i) **Water supply and waste water management**
    a. ULBs are accountable for water supply and quality assurance. Additionally, ULBs are in charge of waste water treatment.

Table 1 Three-step ICCC maturity assessment process.

Stage no.	Description	Maturity assessment process
Stage I	Self-assessment criteria	Cities would be required to submit maturity assessment along with evidence over functional, technical and governance components Only cities which clears minimum threshold score would be eligible for third party audit process Under self-assessment stage, cities would be required to assess and submit assessment over following components of Integrated Command Control Center: **(a)** Functional capability **(b)** Technology readiness **(c)** Governance capability
Stage II	Third-party on-site ICCC maturity assessment	Third party assessment would be required to audit Integrated Command Control Center maturity as per self-assessment submitted by City SPV Third party would be identified and engaged by Ministry of Housing and Urban Affairs, GoI
Stage III	Maturity assessment certification	Cities ICCC would be mapped onto following maturity levels Level 1: Enabled Level 2: Established Level 3: Leader Level 4: Lighthouse This assessment would be done as per defined functional use cases as per city readiness.

b. Solid waste management: This category includes residential garbage collection, collection of construction and demolition debris, waste recycling, and daily waste disposal.

c. Smart street lighting management: This is the management of a network of street lights installed throughout the city limits to ensure pedestrian and motorist safety.

d. Environment: This is a catch-all term for the various sensors installed throughout the facility.

(ii) **City mobility services:** These are city-provided connectivity services that enable citizens to travel between locations. It entails connectivity, accessibility, and the provision of public parking spaces. The three broad domains are as follows:

   a. Transit management (connectivity): This is the management of public transportation vehicles such as buses, taxis, and trains that assist the public in getting around the city.

   b. Transportation planning and control (accessibility): This is the process of planning and controlling transportation services throughout a city in order to manage traffic flow.

   c. Parking solutions for cities: This is the utilization and revenue collection of public parking spaces.

(iii) **Safety and security:** Primarily a police function, this term refers to operations aimed at enhancing public safety and arming police with the surveillance data necessary for both reactive and predictive policing. CCTV surveillance has become a critical component in multiple cities due to the increasing use of video analytics to provide police with timely alerts for action.

(iv) **Crisis management:** These services address major disasters that occur in a city and affect either the entire city (e.g., flooding) or a portion of the city (e.g., fire accident). Medical services, fire departments, and police are all part of a city's crisis management operations, and they may need to respond in unison or in any combination depending on the nature of the emergency. As a critical component of the city's ICCC, crisis management must be established and properly implemented to cover all possible events that could disrupt either a portion or the entire city.

(v) **Convergence:** To support city operations, the city government utilizes certain enterprise systems/applications. GIS services for the city provide users with a comprehensive view of the city by labeling all of a ULB's critical functions on a map.

11. ICCC functional capability assessment

Enterprise resource planning (ERP) is a term that refers to the integrated management of core processes across multiple services that enables the provision of real-time, digitized data about the system. These services span the length and breadth of core services and must therefore be addressed in the city ICCC, as they enable city administrators to visualize data at the pan service and pan city level, with the ability to drill down to specific areas

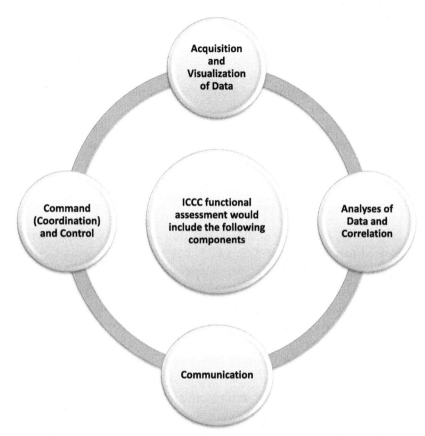

Fig. 4 ICCC functional assessment would include the following components.

of the city or specific services as needed [1]. These services can be evaluated in greater detail using the ICCC platform's functional use case configuration (Fig. 4). In the first stage, cities will self-assess their maturity in specific domains based on their readiness at ICCC.

12. Technology assessment

Technology assessment score would be assigned based on maturity assessment of individual components of ICCC platform to gauge the ICCC capability to support the functional requirements of city administration.

(a) Data acquisition and collection layer includes components for data acquisition and collection from various devices or systems or applications in different formats.

(b) The data analytics and correlation layer consists of components that aggregate and process data for analysis on various dimensions in order to derive intelligence from information gathered from various sources.

(c) Application configuration layer includes components involved in defining and configuring the multiple and complex events and its automated response.

(d) Command and control layer includes components to manage the response, assets, devices, on field users and resources to address civic issues.

13. Governance assessment

A city ICCC will not be able to function to its full potential if it does not have proper governance framework covering people, processes, and policy dimensions to support ICCC operations and its sustenance.

(a) Governance framework: Essentially refers to the presence of governance policies as guidelines for ICCC manpower in terms of nondisclosure agreements, privacy policies, knowledge repositories, and employment policies.

(b) Action-oriented dashboards for city leadership: While it is imperative that an ICCC must be able to display information at aggregated level for city-level management, it is equally important that such dashboards must be regularly utilized by concerned officials for their day-to-day or in crisis situations.

(c) Field force management: This includes but not limited to design and implementation of workforce management plan with well-defined organizational hierarchy, manpower forecasting as well as escalation matrix for the concerned authorities to use and respond to situation in real time.

(d) Resourcing and staffing: With operations of ICCC being a specialized job, recruitment policies must be well documented and approved by all the concerned stakeholders. Once the policies are formulated, it is all the more important the right resources are on-boarded from departments/hired and that they are available round the clock for ICCC operations.

(e) Technical capacity building: The success of a city ICCC depends on the fact that the deployed manpower is regularly trained and retrained on various technological and functional aspects of the ICCC. Also, a training plan must be in place for both staff as well as executives based upon the functions they need to perform.

14. ICCC maturity ranking

Based on scores obtained by cities dynamic ranking could be published in order to instill sense of competitiveness among cities to improve the ICCC as an infrastructure.

(a) City ranking over ICCC
   (1) Overall functional maturity
   (2) Overall technical maturity
   (3) Overall governance maturity
(b) City ranking over ICCC functional maturity components
   (1) Civic utilities
   (2) Mobility services
   (3) Safety and surveillance
   (4) Emergency response
   (5) Convergence
(c) City ranking over ICCC technical components
   (1) Data collection and aggregation capability
   (2) Sop's configuration capability
   (3) Data analytics capability
   (4) Command and control capability
   (5) Data security capability
(d) City ranking over ICCC governance components
   (1) Governance capability
   (2) Field force capability
   (3) Decision-making capability
   (4) Knowledge management capability

15. On-site maturity assessment

On-site maturity assessment is the second stage of our ICCC maturity assessment process. Third-party agency would validate the claims submitted by city to self-assessment stage and gauge the maturity of Integrated Command Control Center. This stage involves more detailed assessment of ICCC capabilities including onsite assessment by third-party auditors over the functional use case implemented by smart cities through Integrated Command and Control Center. Under on-site assessment, qualifying cities are required showcase the ICCC capability by demonstrating functional use cases, which impacts service delivery or in turn livability index of city. Use

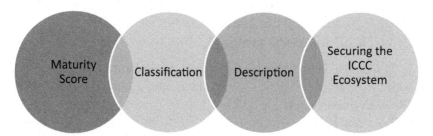

Fig. 5 Identify gaps under design, planning, and implementation for respective of ICCC project under smart cities.

cases are defined to assess the outcome and document the impact on civic operations which are implemented at Integrated Command Control Center. Maturity assessment would also help cities to identify gaps under design, planning, and implementation for respective of ICCC project under smart cities (Fig. 5). Lighthouse cities would emerge as mentors for other cities for guidance and handholding of other cities over managing ICCC.

16. Importance of ICCC security

ICCCs shall be the "nerve center" of a smart city and are envisaged to aggregate information through smart-enabled integrated technologies. The ICCC is expected to provide a comprehensive view of all city operations by enabling monitoring, control, and automation of various functionalities at the individual system level, as well as cross-system analytics. The smart-enabled integrated technologies and devices also bring in their inherent security risks. These technologies and devices connect to ICCC, which poses a threat to ICCC security. Hence, significant increase in the number of interconnected technologies and devices also result in phenomenal increase in the security attack surface. The increase in the security attack surface provides an opportunity to cybercriminals, cyber activists, and nation states to exploit the attack surface to compromise the security of ICCC and, subsequently, of smart city. Hackers and malicious actors now do not need to get direct access to ICCC or data center to compromise systems or applications but can plan attacks through the technologies and devices spread across the smart city. Therefore, as the smart city makes use of the advanced and integrated technology to deliver services to the citizens in an efficient manner, the integrated technology expands the cyber threat landscape. Hence, it becomes imperative to consider the cyber security requirements for a smart city, and particularly of ICCC, in a comprehensive manner.

17. Reasons of increasing the securing of ICCC

Overlooking the security of ICCCs can turn out to be very expensive for an efficient and secure service delivery, and protection of human life. Several serious concerns about the security of smart city services include the following. Through the compromise of the ICCC's integrated traffic management system, kidnappers or malicious actors can monitor live location of the buses, and other parameters and plan their attack accordingly.

- Hacker can add or remove or modify or delete sensitive information from the ICCC database, including residents' personal information, health information and sell the data (personal and health data) same in black market
- State actors from foreign nations can shut down the services (e.g., traffic signals across the city) offered to the citizens and create panic or havoc in the city
- State actors can also use access to post content to spread propaganda and disinformation campaigns
- Organized crimes can be committed by viewing CCTV live feed and then turn off the camera at the time of heist
- GPS systems can be hacked to redirect vehicles such as ambulances, police vans, and school buses leading to chaos in the city
- Aggregation and unauthorized statistical analysis of data collected by ICCCs can be done by miscreants leading to privacy risks and in worst case scenario, loss of human life
- What should be done?

Security and privacy should be considered across all phases of an ICCC development, design, implementation, operations along with preparation for long-term assurance.

(a) **Cyber security framework and security by design**

    (i) A cyber security framework should be developed aimed at building a secure and resilient ICCC for citizens and stakeholders of smart city. The ICCC is expected to provide a comprehensive view of all city operations by enabling monitoring, control, and automation of various functionalities at the individual system level, as well as cross-system analytics.

        ■ MoHUA guidelines vide circular K-15016/61/2016-SC-1, dated May 20, 2016

        ■ Government of India guidelines on Data Security

- IT Act and Amendment 2008
- CERT-IN guidelines
- CMP guidelines on countering cyber attacks
- International standards including ISO 27001, NIST Cyber Security Framework

(ii) A secure network architecture should be designed following a layered security approach. Security solutions as detailed out in MoHUA guidelines should be considered, as appropriate, to protect the ICCC.

(iii) Cyber security awareness trainings should be provided to different focus groups responsible for the security of ICCC.

(b) Security while implementation

A governance mechanism should be setup to ensure that ICCC implementation conforms to secure requirements. Security assessment should be performed for all the ICCC associated applications, systems, and devices before Go-Live.

(c) Security during operations

Secure procedures should be followed during the ICCC operations. All the changes, operations, and monitoring of ICCC applications and systems should be performed in a controlled manner following a well-defined process. A security operations center should be setup comprising of a threat analytics solution to give a reasonable security assurance for ICCC from emerging cyber threats. An incident response mechanism should be setup to respond in a coordinated manner to any security attack.

(d) Security assurance

A regular process should be setup to assess the ICCC compliance to security and regulatory requirements on a regular basis. The gaps identified during the assessment should be auctioned for mitigation depending upon the criticality.

18. Conclusion

Thus, it is learnt that ICCCs are like the nervous system to the body of a smart city. It pictures with aggregate information through integrated technologies. It provides a detailed view of operations of all the cities by monitoring and controlling various functions. These functions are classified as individual and system level where there is a risk of security which has to be addressed. Hence, there should be a significant increase in the security of ICCC from security attacks through more interconnected devices with smart technologies [1]. It should be noted that ICCC should be protected with advanced and integrated smart technology while put into use by the public in a very effective and efficient manner. The efficient use of integrated technology is based on the ability to extract the required data which has to acquire from different layers of the ICCC management. It assists the ICCC in handling the real-time events and provides intelligence and information from various sources.

The ability to make effective decision on the operations of the smart city through optimization is the real value of ICCC. In many cities, the implementation models adopted were different because they have decided to set up their own ICCC with unique delivery of particular functions. These cities use separate hub and unique integrated models for various functions and controls the city operations including the daily scenarios. Benchmarking the ICCC in smart cities across the country is possible through evaluation model. It has to be done along with standards and analytics to gather information on the performance and to optimize the governance and integrated technology that help to make smart cities more developed. Also, the ICCC investment should be done appropriately for necessary features. Further the area of improvements should be identified and appropriate operations should be implemented. Thus, the assessment of smart cities is very important to understand the impact of ICCC on various parameters and to identify the operations that have to be improved by managing various operations.

Reference

[1] Ministry of Housing & Urban Affairs (MoHUA), Integrated Command and Control Center Maturity Assessment Framework and Toolkit Maturity Assessment Framework and Toolkit to Unlock the Potential of Integrated Command and Control Centers (ICCCs), 2018 (December).

PART FOUR

Smart environment

CHAPTER NINE

Renewable energy based hybrid power quality compensator based on deep learning network for smart cities

Ginnes K. John, M.R. Sindhu, and T.N.P. Nambiar
Department of Electrical and Electronics Engineering, Amrita School of Engineering, Amrita Vishwa Vidyapeetham, Coimbatore, India

1. Introduction

Recent advancements in the integration of electric vehicles and distributed power generators with the grid, as well as the use of solid-state controllers in industries and commercial centers, have resulted in unpredictable variations in load demand, power generation, and harmonic injection into the grid. These factors cause power quality to deteriorate, resulting in voltage sag/swell, waveform distortion, fluctuation, flickering, and unbalance, among other things. This has an impact on the reliability, security, and stability of vital and sensitive equipment used in data centers, hospitals, banking services, and other industries [1,2]. Compensators such as passive filters, active filters, hybrid filters, dynamic voltage restorers (DVR), STATCOM, and others were installed to reduce these power quality disturbances [3]. Typically, these devices are placed near certain load terminals that demand higher power quality based on load requirements. In modern power systems, a microgrid is used to feed a group of electrical loads such as smart cities.

The US Department of Energy defines microgrid as "A group of interconnected loads and distributed energy resources within clearly defined electrical boundaries that acts as a single controllable entity with respect to the grid. A microgrid can connect and disconnect from the grid to enable it to operate in both grid-connected or island mode" [4].

The Consortium for Electric Reliability Technology Solutions (CERTS) defines a microgrid as "The MicroGrid concepts assumes an aggregation of loads and micro sources operating as a single system providing both power and heat. The majority of the micro sources must be power electronic based to provide the required flexibility to ensure operation as a single aggregated system. This control flexibility allows the CERTs Micro-Grid to present itself to bulk power system as a single control unit that meets the local needs for reliability and security" [5].

Microgrids comprise energy storage systems, load control, and power quality enhancement technology in addition to renewable energy sources to provide stable power supply, minimize losses, and provide good quality to users. Solar PV systems, wind turbines, fuel cells, and bio-fueled micro turbines all are examples of renewable generation. Energy storage solutions such as batteries and ultracapacitors are preferred.

The IEEE P1547 standard is the most widely adopted for connecting distributed sources to the interconnected power system [6]. In a microgrid, these capabilities are assured by adding peer-to-peer architecture and a plug-and-play approach for each component [5]. Because distribution resources are local power supplies, they enhance system efficiency by lowering carbon emissions, reducing losses, and maximizing the use of power resources. On the other hand, they can lead to problems like meshed power flow, bidirectional power flow, increased fault levels, a lack of protection system coordination, low power factor, power frequency variations, voltage sag/swell/flicker, harmonics, and so on. Control, interface, and protection of each micro resource, as well as control of power flow, load sharing amongst resources, voltage regulation, and system stability, are all included in the microgrid controller [7].

A benchmark of low-voltage microgrid network including multifeeder microgrid network was proposed by Stavros Papathanassiou at CIGRE Symposium held at Athens, Greece, in the year 2005 [8]. In this benchmark of LV feeder, general characteristics of the network, description of the benchmark low-voltage feeder, and consumer demand characteristics are summarized. This research study presents a benchmark LV microgrid network that can be used for both steady-state and transient simulations [8].

The benchmark LV microgrid network has PV systems, wind turbines, microturbines (CHP generation), and fuel cells, and its sizes and relevant installation locations are indicated. The total installed capacity of the microsources is around 2/3 of the feeder's maximum load demand, allowing for load management simulation. A fast-responding central storage device, such as a flywheel, batteries, or ultra-capacitors, is utilized to enable the microgrid's islanded operation [8]. To provide decentralized active power/frequency concerted regulation, a centralized control approach is necessary [9].

The benchmark network preserves essential technical characteristics of real-world utility grids while removing the complexity of actual networks,

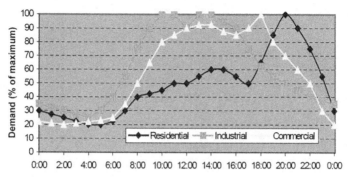

Fig. 1 Daily load curves for the benchmark LV networks: three load types. *(No permission required.)*

allowing for efficient microgrid modeling and simulation [8]. The daily load curves of LV benchmark network are shown in Fig. 1 that the time at which the maximum load demand of these loads is also different. The loads are considered as three types, namely, residential, industrial, and commercial, and behavior of these loads can be linear as well as nonlinear [8].

2. CIGRE LV multifeeder microgrid: Analysis of power quality issues

The load demands of a CIGRE LV multifeeder microgrid are taken to suite the demands of the case study as shown in Fig. 2. In this case study, load LA (balanced linear load, unbalanced nonlinear load) and load LA and LAAA (unbalanced linear load, critical, and sensitive load) are considered. The characteristics of these loads are mentioned in Table 1. Fig. 2 shows a single line diagram of CIGRE LV multifeeder microgrid that does not have any compensation devices. Generally, smart city loads are fed by two feeders: a preferred feeder (PF) and an alternate feeder (AF), both of which are of standard conventional grade power quality. It is made up of three feeders. Different degrees of power quality are required for various feeder loads: LA (residential feeder), LAA (industrial feeder), and LAAA (commercial feeder). The feeder load LA may tolerate problems such as harmonics, interruption, low power factor, sag/swell, and so on. The feeder load LAA has critical and sensitive loads that require both good quality of voltage and current. The feeder load LAAA demands only good quality of voltage at its feeder terminals. The behavior of loads is investigated using MATLAB simulation. Each load creates harmonics, a decrease in power factor, and imbalance. These problems will propagate across the system.

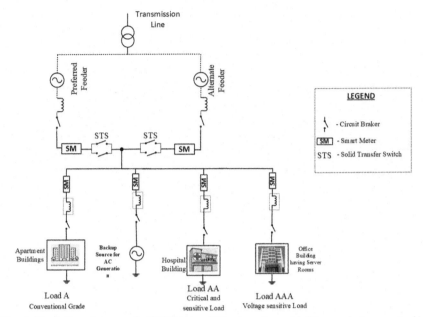

Fig. 2 Single line diagram of a CIGRE LV multifeeder microgrid. *(No permission required.)*

Table 1 Characteristics of load.

Load	Behavior of loads	Time instant of switching ON(s)	kVA Rating of load
Load LA (2 kVA)	Balanced linear load	0.0	0.5
	Unbalanced nonlinear load (1.5 kVA)	0.1	0.5 (balanced nonlinear)
		0.06	1 (unbalanced linear load)
Load LAA and Load LAAA (1 kVA)	Unbalanced linear load	0.06	0.75 (unbalanced linear load)
	Critical and sensitive load	0.14	0.25 (balanced nonlinear load)

Case studies are done as two parts, namely, case 1—without any compensation and case 2—with compensation.

2.1 Case 1: Without compensation

In case 1, a CIGRE LV multifeeder microgrid without any compensation is discussed as shown in Fig. 2. Time instants of switching of loads are shown in Table 1. The deterioration in power quality at feeder terminals is depicted as waveforms in Fig. 3, and the relevant values are listed in Table 2.

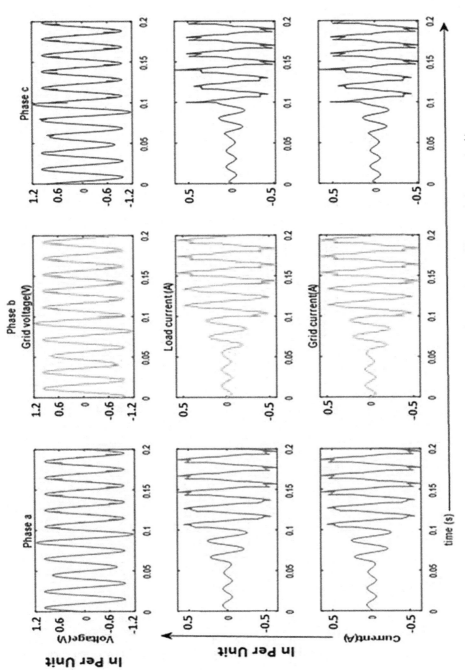

Fig. 3 Three-phase grid voltages, load currents, and grid currents waveforms—Case I. *(No permission required.)*

Table 2 System voltage, current, power factor, and TDD measurements—Case I.

Time interval	0–0.04 s	0.04–0.06 s	0.06–0.08 s			0.08–0.1 s			0.1–0.14 s			0.14–0.2 s		
Phase parameters	A, B & C	A, B & C	A	B	C	A	B	C	A	B	C	A	B	C
Grid/load voltage (rms) (p.u.)	1	0.8	1			1.2			1			1		
Grid/load power factor	0.7	0.7	0.6	0.6	0.6	0.6	0.6	0.6	0.68	0.6	0.6	0.7	0.7	0.7
Grid/load current TDD (%)	0.03	0.03	0.95	0.84	1.69	0.95	0.84	1.69	18.9	21.5	23.6	23.2	25.6	28.5

A balanced linear load of 0.5 kVA is connected from time $t=0$ to time $t=0.06$ s. The grid suffers from voltage sag between $t=0.04$ and $t=0.06$ s. The grid is affected by voltage swell from $t=0.08$ to $t=0.1$ s. Balanced nonlinear load is connected with other linear loads within the time interval $t=0.1$–0.14 s, and the grid current total demand distortions (TDDs) are 18.9%, 21.5%, and 23.06%, respectively. Critical and sensitive loads are connected with all other loads from $t=0.14$ to 0.2 s and the corresponding TDDs for A, B, and C phases are 23.29%, 25.63%, and 28.58%, respectively.

According to IEEE-519 standards, the quality of electricity supplied is too poor [10]. The foregoing study shows that the system has sag, swell, imbalance, a low power factor, and high current harmonics.

The deregulation of electric power energy has increased public awareness of power quality among various user groups [11]. In various publications, the subject of power quality and its problems relating to the electric power network have been discussed [11–14].

2.2 Custom power devices

Custom power is a term that refers to the employment of power electronics controller devices, known as custom power devices, in power distribution systems to provide a quality of power to sensitive customers, particularly industrial and commercial consumers [11]. Due to the self-supporting of dc bus voltage with a large dc capacitor, VSI is frequently employed for the development of custom power devices. The custom power devices are generally divided into network reconfiguring type and compensating type [11].

GTO or thyristor-based network reconfigurable custom power devices are commonly used for quick current limiting and current breaking. Solid-state current limiter, static transfer switch, and static breaker are the most common network reconfiguration custom power devices. In the fault circuit, the static current limiter device places a limiting inductor and removes the inductor from the fault circuit once the fault has been cleared. A static transfer switch is made up of two thyristor or GTO blocks that are connected in parallel. The transfer switching time varies between 0.25 and 0.5 cycles of the fundamental frequency. The fundamental advantage of static transfer switch is that it continually conducts the load current. That is, it provided customers with uninterrupted power at the distribution level. The load current leads the conducting losses in high-power applications, which is a drawback of this switch. Conducting losses are typically between 0.5% and 1% of the load power. Solid-state breaker is a high-speed switching mechanism used to prevent electrical faults and safeguard distribution systems from excessive currents. The breaker's voltage and current rating define the number of switching devices required, the cost, and the breaker's losses. It has an autoclosing feature.

Mostly preferred solutions to the power quality issues are passive filters and active filters [14]. Passive filters are advantageous since they have a low starting cost and a high efficiency. However, it has a number of limitations, including instability, fixed compensation, resonance with both supply and loads, and utility impedance. Active power filters have been employed to overcome the limitations of passive filters. There are three types of active power filters: shunt, series, and hybrid. Shunt APF is used to compensate for current-based distortions, whereas series APF is used to compensate for voltage-based distortions. Hybrid APF is a combo of series and shunt types that are used for filtering high order harmonics and in common applications; their rating is sometimes quite close to load (up to 80% load). Active filtering, load balancing, power factor enhancement, and voltage regulation (sag/swell) are all possible with compensatory custom power devices. There are three varieties of these devices: static shunt compensator, series compensator, and hybrid compensator. These are also referred to as DSTATCOM, DVR, and UPQC [2,11,13].

The distribution static compensator (DSTATCOM) is a static compensator device (STATCOM, FACTS controller) based on a voltage source inverter (VSI). It is used to keep bus voltage sags at the required level by delivering or receiving reactive power in the distribution system. A VSI, a dc energy storage device, an ac filter, and a coupling transformer constitute the DSTATCOM. VSI converts DC voltage into controllable ac voltage in the power circuit, which is then synchronized by an ac filter and connected to the AC distribution line via a coupling transformer. DSTATCOM's working concept, which involves continuous monitoring of load voltages and currents, determines the amount of compensation required by the distribution system in the event of a variety of disturbances. The angle between the ac system and VSI voltages controls active power flow; the difference between the magnitudes of these voltages controls reactive power flow. The DSTATCOM is capable of both current and voltage control [2,15,16].

DVR is a series-connected custom power device that uses a coupling transformer to inject a dynamically regulated voltage in magnitude and phase into the distribution line, correcting load voltage. It comprises a series connection of an energy storage device, a dc-dc converter, voltage source inverter, ac filter, and coupling transformer. A compensatory voltage is generated by the inverter and introduced into the distribution system via a series matching transformer. The DVR controllers create a reference voltage, compare it to the source voltage, and inject synchronized voltage to keep the load voltage constant in the case of voltage irregulation. To synchronize injected voltage, the energy storage device delivers the needed power. The ac filter compensates for the impacts of coupling transformer winding and switching losses in control signal generation strategies for VSI [2,14,15].

The Unified Power Quality Compensator (UPQC) is a back-to-back connection of shunt and series compensators via a common dc bus voltage. The dc link storage capacitor is attached between two voltage source inverters and operates as a shunt and series compensator while in use. It is a very adaptable device that can simultaneously restrict current in a shunt and voltage in a series. It can simultaneously balance the terminal voltage and eliminate negative sequence current components [2].

2.3 Custom power park

At present scenario, more and more penetration of renewable systems and inclusion of nonlinear system creates power quality issues. According to CIGRE standard, loads are categorized as industrial, commercial, and residential loads. These loads have its own load profile patterns and subjected to power quality issues. Industrial loads such as Industries process control, robotic tools, etc., are affected more and cause huge financial losses. These industries will pay more money to get good quality of power. A custom power park (CPP) will supply good quality of electrical power with the help of custom power devices (CPDs) [3,12,17,18]. In this era of huge number of renewable energy resources, such as solar, wind, and their integration through power converters, integration of electric vehicles (EV) with grid to enable bidirectional power transfer and quality of power available to customers is a serious concern [2]. Excess penetration of renewable energy unscheduled charging/discharging of EV batteries may cause PQ issues such as voltage swell, voltage sag, under voltage, over voltage, interruption, harmonics, imbalance, excess Q burden. Based on the category of customers, quality of power is selected in CPP. A CPP can get maximum financial benefit of it and can provide best quality of power with operation and control of minimum number of custom power devices (CPDs), minimum number of CPDs that can be configured and controlled to provide series (voltage) compensation, shunt (current) compensation, voltage and current harmonics, real power, reactive power, or combination of these. Whenever a large number of custom power devices to be controlled in a custom power park, older learning algorithms like supervisory control of discrete event system (SCDES) [19] are no longer performing well. So a custom power park controlled by deep learning network is proposed in this work.

2.4 Deep learning network

The use of smart grid technology necessitated the development of appropriate monitoring, classification, and control techniques to ensure the system's success. Modern grids incorporate distributed renewable resources such as solar and wind, as well as energy storage schemes, power quality enhancement technologies, and monitoring, communication, and control mechanisms. The integration of various system components, as well as measuring devices and control systems, improves the power grid's performance significantly. Various power quality concerns, such as voltage swell, voltage sag, harmonics, poor power factor, imbalance, and power oscillations, may affect the smart power grid [14]. As a result, the smart grid's effectiveness is determined by the effectiveness of data sensing, classification, and control mechanisms. Data on system health must be collected from each and every component in the smart grid. The management and analysis of big data secured from the system are key to the control technique's success. Big data analysis techniques have been successfully utilized in areas such as solar power generation forecasts, grid price, load demand, fault detection, and classification in the power system [19]. Electrical quantities were regularly measured and compiled in traditional power systems. Technical professionals or signal processing techniques analyze these data. These methods are incapable of handling the large amounts of data generated by today's electricity infrastructure. As a result, technologies for self-processing automated data analysis are required.

Artificial intelligence-based techniques such as artificial neural network ANN, expert systems, support vector machines were preferred by researchers in early stages of automated analysis [20]. They were applied to power quality analysis and event classification also. But the quality of the result certainly depended on the signal processing techniques chosen for the work. ANNs were useful while dealing with nonlinear data, which utilizes activation functions such as sigmoid, hyperbolic tangent, etc., for this nonlinear transformation. Later, machine learning-based techniques used kernel functions as polynomial kernel, sigmoid kernel, Gaussian kernel, etc., for mapping the raw data to real application domain [20,21].

Commonly used machine learning techniques are support vector machine (SVM), principal component analysis (PCA), linear discriminant analysis (LDA), etc. These machine learning methods have many drawbacks as they can handle only small datasets, overfitting or underfitting of data, applies supervised learning, and does not ensure optimal hyperparameters. Here technical expert selects the network features, and the selected signal processing technique extracts the features. While dealing with big data in modern power grid, automatic feature extraction methods are needed, as human expert cannot perform quality analysis of huge data involved.

Deep learning (DL) methods automatically extract features of the data. Here, in each layer, nonlinear transformation techniques such as convolution sigmoid, ReLu, etc., are applied. A combination of variational autoencoders, generative adversial network, and dropout techniques such as max-pooling and upsampling methods improves the capability of DL [19,20,22].

3. Renewable energy-based hybrid power quality compensator with CIGRE LV multifeeder microgrid

The ReHPQC as shown in Fig. 4 is equipped with a centralized controller based on deep learning network, which aids in the maintenance of required power quality at feeder load terminals LAA and LAAA. The current and voltage signals of each feeder loads are fed back to the centralized controller. This controller maintains the power quality at the feeder load terminals LAA and LAAA with the help of the selected control strategy. Through DC/DC and DC/AC converters, solar PV panels (PV) and battery energy storage systems (BS) are incorporated into the microgrid. As a result of this extra capability in ReHPQC, the microgrid can operate more reliably and at a reduced cost. ReHPQC employs the PWM control algorithm for series/voltage compensation and *icosφ* algorithm for shunt/current compensation [16].

The CIGRE microgrid's power quality is analyzed using MATLAB/ SIMULINK, and the key power quality issues include (i) voltage sag/ swell/harmonics/ unbalance, (ii) poor power factor, (iii) current harmonics/unbalance. The input data include solar irradiation and power generation, ambient and PV module temperatures, SoC of BS, voltages and currents at feeders, DG, PV, and load terminals. To account for seasonal dependency on power generation and loads, these data samples are gathered for one year at 20 min intervals and categorized according to IEEE 1159-2019 standards.

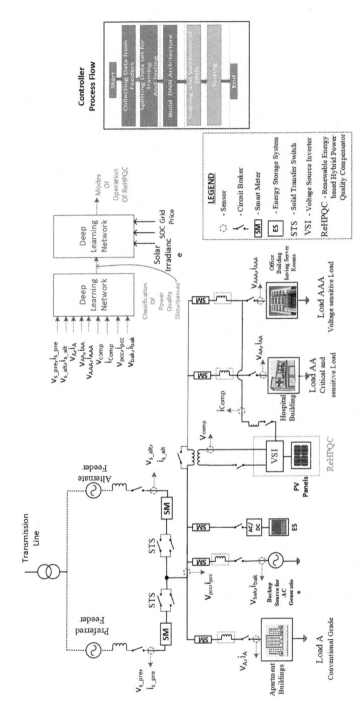

Fig. 4 ReHPQC in CIGRE multifeeder microgrid. *(No permission required.)*

For each of the following cases, a few samples are collected: (a) normal voltage, (b) voltage sag, (c) voltage imbalance, (d) voltage harmonics, (e) normal voltage + harmonics, (f) voltage sag + harmonics, (g) voltage swell + harmonics, (h) voltage imbalance + harmonics, (i) reactive power demand, (j) voltage sag + reactive power demand (k) voltage swell with reactive power demand, (l) voltage imbalance with reactive power demand, (m) voltage harmonics + reactive power demand, (n) normal voltage + harmonics + reactive demand, (o) voltage sag + harmonics + reactive power demand, (p) voltage swell + harmonics + reactive power demand, (q) voltage imbalance + harmonics + reactive power demand.

3.1 Working of deep neural network used in ReHPQC in CIGRE multifeeder microgrid

The following signals can be sensed, as shown in Fig. 4, namely, preferred feeder voltage (vs_pre), preferred feeder voltage (is_pre), alternate feeder voltage (vs_alt), alternate feeder current (is_alt), load terminal voltage A (v_A), load terminal current A (i_A), load terminal voltage AA (v_{AA}), load terminal current AA(i_{AA}), load terminal voltage AAA (v_{AAA}), load terminal current AA(i_{AAA}), compensator voltage (v_{comp}), compensator current (i_{comp}), PCC voltage (vpcc), PCC current (ipcc), backup voltage (v_{bak}), and backup current (i_{bak}).

Even though the above-mentioned signals can be sensed, only four signals, namely, line voltages at PCC and phase current A at the PCC are considered for the case study. For each of the four signals, 1500 samples are taken for the analysis. The classification classes are voltage sag, voltage swell, and current harmonics.

The data are processed using the MATLAB with a 70:15:15 split for training, validation, and testing, respectively. A deep neural network (DNN) controls the ReHPQC. Fig. 5 shows the DNN network used in the MATLAB-based simulation. It has 3 hidden layers that has 10 neurons in each hidden layer and one output layer. It is being feed by 4 input signals and provide 3 outputs signals.

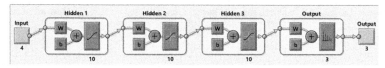

Fig. 5 DNN network used for the analysis. *(No permission required.)*

Following algorithms may be used as multilayer neural network training function in MATLAB, namely, scaled conjugate gradient backpropagation, Levenberg–Marquardt, BFGS Quasi-Newton, resilient backpropagation, conjugate gradient with Powell/Beale Restarts, Fletcher–Powell conjugate gradient, and one-step secant.

The cross entropy performance function is used for all the above-mentioned algorithms in the analysis made here.

The following parameters are being used to evaluate an algorithm's performance.

- *Accuracy*: It offers the model's overall accuracy, which is the percentage of total samples correctly identified by the classifier.
- *Misclassification rate*: It shows what percentage of predictions was wrong. It is also referred to as a classification error.
- *Precision*: It shows what percentage of positive predictions was truly positive.
- *Recall*: It represents the percentage of all positive samples the classifier accurately predicted as positive.
- The confusion matrix acquired from the simulation can be used to compute these values. Simulation yields four confusion matrices: training confusion matrix, validation confusion matrix, test confusion matrix, and all confusion matrices.

The result with scaled conjugate gradient backpropagation algorithm is shown below.

3.2 Scaled conjugate gradient backpropagation

As previously stated, the following three classes are taken into account: class 1—sag, class 2—swell, and class 3—current harmonics.

Table 3 provides parameters obtained from training confusion matrix data.

Parameters obtained from validation confusion matrix data are given in Table 4.

Parameters obtained from test confusion matrix data are given in Table 5.

Table 3 Parameters obtained from training confusion matrix.

Class	Precision	Recall	Miscalculation rate	Overall accuracy
Sag	0.94	0.94	0.077	0.923
Swell	0.935	0.879		
Current harmonics	.897	0.946		

Table 4 Parameters obtained from validation confusion matrix data.

Class	Precision	Recall	Miscalculation rate	Overall accuracy
Sag	1	0.8	0.087	0.913
Swell	0.875	1		
Current harmonics	0.857	1		

Table 5 Parameters obtained from test confusion matrix data.

Class	Precision	Recall	Miscalculation rate	Overall accuracy
Sag	0.833	0.833	0.13	0.87
Swell	0.1	0.9		
Current harmonics	0.75	0.857		

Table 6 Parameters obtained from all confusion matrix data.

Class	Precision	Recall	Miscalculation rate	Overall accuracy
Sag	0.938	0.9	0.087	0.913
Swell	0.938	0.9		
Current harmonics	0.87	0.9		

Table 7 All confusion matrices.

	Sag	450	0	30
	Swell	30	450	0
Output class	Current harmonics	20	50	470
		Sag	Swell	Current harmonics
		Target class		

Parameters obtained from all confusion matrix data provided in Table 6. All confusion matrices are provided in Table 7.

Fig. 6 shows ROC curve (receiver operating characteristic curve) which is a plot that summarizes a binary classification model's performance on the positive class. The false positive rate is shown on the x-axis, while the true positive rate is shown in the y-axis. A curve that runs from the bottom left to the top right and bows toward the top left can be formed by analyzing the true positive and false positives for different threshold levels. The top left of the plot represents the best possible classifier that achieves perfect competence (coordinates 0,1). The class 2 (swell), is the best classification followed by class 3 (current harmonics) and class 1 (sag).

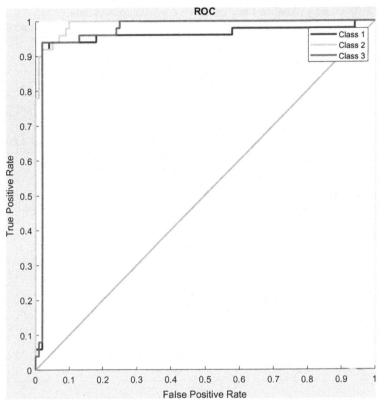

Fig. 6 ROC curve using scaled conjugate gradient backpropagation algorithm. *(No permission required.)*

Similarly, parameters of all algorithms can be calculated. Due to page constraints, only the summary of parameters of overall accuracy obtained from all confusion matrix data is shown for all algorithms tested (Table 8).

After comparing the deep neural network model to the learning algorithms offered, it was learnt that the Levenberg-Marquardt method performs the best. One-step secant has the worst result.

After the classification of power quality disturbances, the selection of operation of the switch of the custom power devices to be ON is decided by the another DNN network that have the input of state of charge of battery, solar irradiance level, and grid price. ReHPQC will act as a series compensator for the sag and swell classification classes, and as a shunt compensator for the current harmonics classification class as shown in Table 9.

Table 8 Overall accuracy of each algorithm.

Algorithms	Overall accuracy
Scaled conjugate gradient backpropagation	0.913
BFGS Quasi-Newton	0.932
Resilient backpropagation	0.92
Levenberg-Marquardt	0.94
Conjugate gradient with Powell/Beale restarts	0.931
Fletcher-Powell conjugate gradient	0.91
One-step secant	0.87

Table 9 Current harmonics classification.

Classification	ReHPQC works as
Sag	Series compensator
Swell	Series compensator
Current harmonics	Shunt compensator

3.3 Case 2—With compensation using ReHPQC

Adding custom power devices to Fig. 2 gives a renewable energy-based hybrid power quality compensator (ReHPQC)-based CIGRE LV multifeeder microgrid, displayed in Fig. 4, which is used for case study in this paper. In this case, the compensations from ReHPQC are provided to the load terminals. Time instants of switching of loads are same as in case 1 and case 2.

In simulation analysis, it is shown that sag affects from 0.04 s to 0.06 s and swell from 0.08 s to 0.1 s. During the sag and swell, ReHPQC provides compensating voltages at the load terminal AAA and obtains normal voltage values, as illustrated in Fig. 7.

The system needed current compensation at the load terminal AA from $t = 0.1$ to 0.2s since the critical and sensitive loads were connected during this time. This time ReHPQC acts as shunt active filter to compensate reactive and distortion power at load terminal AA. The power factor and TDD at PCC are improved a lot compared to case 1 as shown in Table 4 and thus improving current quality at PCC.

Fig. 7 depicts the related waveforms, and Tables 10 and 11 list the pertinent parameters.

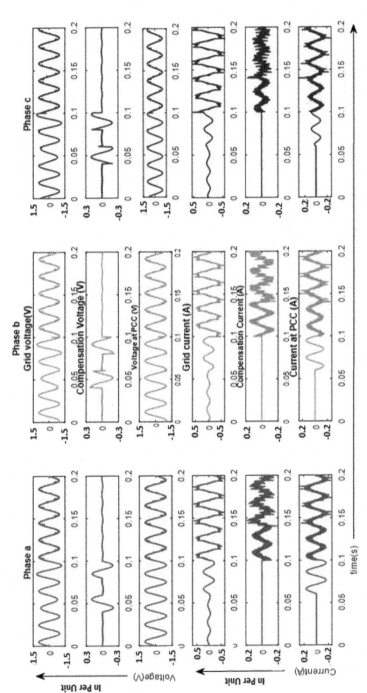

Fig. 7 Three-phase grid voltage, compensator voltage, voltage at PCC, grid current, compensator current, current at PCC with ReHPQC compensation. *(No permission required.)*

Table 10 The pertinent parameters for time interval 0–0.1 s.

Time interval	0–0.04 s	0.04–0.06 s	0.06–0.08 s	0.08–0.1 s
Phase parameters	A, B & C	A, B & C	A, B & C	A, B & C
Grid voltage (rms) (p.u.)	1	0.8	1	1.2
Voltage THD (%)	0.03%	0.2%	0.3%	0.2%
Compensation voltage (rms) (p.u.)	0	0.2	0	−0.2
Voltage at PCC (rms) (p.u.)	1	1	1	1

Table 11 The pertinent parameters for time interval 0.1–0.2 s.

Time interval	0.1–0.14 s			0.14–0.2 s		
Phase parameters	A	B	C	A	B	C
PCC voltage (rms) (p.u.)	1			1		
Power factor at PCC	0.99	0.99	0.99	0.99	0.99	0.99
TDD at PCC (%)	8.01	8.87	8.97	8.18	9.17	9.23
Grid current TDD (%)	12.1	13.2	18.1	16.4	19.2	21.1

4. Conclusion

Recent developments using static power semiconductor switches led to the power quality issues in the smart grid electrical equipment's especially in the critical and sensitive loads. This paper looks at case studies of a renewable energy-based hybrid power quality compensator (ReHPQC) in a CIGRE multifeeder microgrid. The CIGRE microgrid's power quality is analyzed using MATLAB/SIMULINK, and the key power quality issues considered for the simulation studies are voltage sag, swell, and current harmonics. The classification and control models were trained, validated, and tested, and a dual-stage deep learning network was created for ReHPQC coordination and control. According to the severity of power quality issues, the DL controller creates switching signals for ReHPQC. A ReHPQC can configure to serve users with different levels of power quality.

References

[1] A.B. Nassif, Y. Wang, Iraj Rahimi Pordanjani, Power Quality Characteristics and Electromagnetic Compatibility of Modern Data Centres, in: IEEE Canadian Conference on Electrical & Computer Engineering (CCECE), 2018.

[2] L. Wang, Z. Qin, T. Slangen, P. Bauer, T. van Wijk, Grid impact of electric vehicle fast charging stations: trends, standards, issues and mitigation measures—an overview, IEEE Open J. Power Electron. 2 (2021) 56–74, https://doi.org/10.1109/ojpel.2021.3054601.

[3] A. Ghosh, G. Ledwich, Power Quality Enhancement Using Custom Power Devices, Springer US, 2002, https://doi.org/10.1007/978-1-4615-1153-3.

[4] D.T. Ton, M.A. Smith, The U.S. Department of Energy's Microgrid Initiative, Electricity J. 25 (8) (2012) 84–94, https://doi.org/10.1016/j.tej.2012.09.013.

[5] R. Lasseter, A. Akhil, C. Marnay, J. Stephens, J. Dagle, R. Guttromson, et al., Integration of Distributed Energy Resources. The CERTS Microgrid Concept (No. LBNL-50829), Lawrence Berkeley National Lab. (LBNL), Berkeley, CA, 2002.

[6] T.S. Basso, R.D. DeBlasio, IEEE P1547-series of standards for interconnection, in: Proceedings of the IEEE Power Engineering Society Transmission and Distribution Conference, vol. 2, 2003, pp. 556–561.

[7] T.E. Hoff, H.J. Wenger, B.K. Farmer, Distributed generation: an alternative to electric utility investments in system capacity, Energy Policy 24 (2) (1996) 137–147, https://doi.org/10.1016/0301-4215(95)00152-2.

[8] S. Papathanassiou, N. Hatziargyriou, K. Strunz, Power systems with dispersed generation: technologies, impacts on development, operation and performances, in: CIGRE Symposium, 2005.

[9] D. Georgakis, S. Papathanassiou, N. Hatziargyriou, A. Engler, C. Hardt, Operation of a prototype microgrid system based on micro-sources equipped with fast-acting power electronics interfaces, in: PESC Record—IEEE Annual Power Electronics Specialists Conference, vol. 4, 2004, pp. 2521–2526, https://doi.org/10.1109/PESC.2004.1355225.

[10] IEEE Standard, IEEE recommended practice and requirements for harmonic control in electric power systems, in: IEEE Standard, 519, 2014.

[11] S. Gupt, A. Dixit, N. Mishra, S.P. Singh, Custom power devices for power quality improvement: a review, Int. J. Res. Eng. Appl. Sci. 2 (2) (2012) 1646–1659.

[12] A. Kharrazi, Y. Mishra, V. Sreeram, Discrete-event systems supervisory control for a custom power park, IEEE Trans. Smart Grid 10 (1) (2017) 483–492.

[13] R.K. Majji, J.P. Mishra, A.A. Dongre, Optimal switching control of series custom power device for voltage quality improvement, in: 3rd International Conference on Energy, Power and Environment: Towards Clean Energy Technologies, ICEPE 2020, Institute of Electrical and Electronics Engineers Inc., 2021, https://doi.org/10.1109/ICEPE50861.2021.9404437.

[14] F. Nejabatkhah, Y.W. Li, H. Tian, Power quality control of smart hybrid AC/DC microgrids: an overview, IEEE Access 7 (2019) 52295–52318, https://doi.org/10.1109/access.2019.2912376.

[15] A.A. Alkahtani, S.T.Y. Alfalahi, A.A. Athamneh, A.Q. Al-Shetwi, M.B. Mansor, M.A. Hannan, V.G. Agelidis, Power quality in microgrids including supraharmonics: issues, standards, and mitigations, IEEE Access 8 (2020) 127104–127122, https://doi.org/10.1109/access.2020.3008042.

[16] G.K. John, M.R. Sindhu, T.N.P. Nambiar, in: Hybrid VSI Compensator for AC/DC Microgrid, Proceedings of 2019 IEEE Region 10 Symposium, TENSYMP 2019, Institute of Electrical and Electronics Engineers Inc., 2019, pp. 738–743, https://doi.org/10.1109/TENSYMP46218.2019.8971222.

[17] A. Ghosh, Power quality enhanced operation and control of a microgrid based custom power park, in: IEEE International Conference on Control & Automation (ICCA'09), 2010.

[18] A. Ghosh, A. Joshi, A new approach to load balancing and power factor correction in power distribution system, IEEE Trans. Power Deliv. 15 (1) (2000) 417–422, https://doi.org/10.1109/61.847283.

[19] W. Qiu, Q. Tang, J. Liu, W. Yao, An automatic identification framework for complex power quality disturbances based on multifusion convolutional neural network, IEEE Trans. Ind. Inform. 16 (5) (2020) 3233–3241, https://doi.org/10.1109/TII.2019.2920689.

[20] S. Wang, H. Chen, A novel deep learning method for the classification of power quality disturbances using deep convolutional neural network, Appl. Energy 235 (2019) 1126–1140, https://doi.org/10.1016/j.apenergy.2018.09.160.

[21] N. Mohan, K.P. Soman, R. Vinayakumar, Deep power: deep learning architectures for power quality disturbances classification, in: Proceedings of 2017 IEEE International Conference on Technological Advancements in Power and Energy: Exploring Energy Solutions for an Intelligent Power Grid, TAP Energy 2017, Institute of Electrical and Electronics Engineers Inc., 2018, pp. 1–6, https://doi.org/10.1109/TAPENERGY. 2017.8397249.

[22] J. Ma, J. Zhang, L. Xiao, K. Chen, J. Wu, Classification of power quality disturbances via deep learning, IETE Tech. Rev. 34 (4) (2017) 408–415, https://doi.org/10.1080/ 02564602.2016.1196620.

CHAPTER TEN

Predicting subgrade and subbase California bearing ratio (CBR) failure at Calabar-Itu highway using AI (GP, ANN, and EPR) techniques for effective maintenance

Kennedy C. Onyelowe[a,b], John S. Effiong[b], and Ahmed M. Ebid[c]
[a]Department of Civil and Mechanical Engineering, Kampala International University, Kampala, Uganda
[b]Department of Civil Engineering, Michael Okpara University of Agriculture, Umudike, Nigeria
[c]Department of Structure Engineering & Construction Management, Faculty of Engineering, Future University in Egypt, New Cairo, Egypt

1. Overview

Road pavement failure is a common phenomenon in the developing countries especially in Nigeria, where its construction is not done by strictly following international standard and design consideration in terms of materials selection. It has been observed that flexible pavements, which are the commonest pavements in Nigeria, fail few weeks and in some cases few months after construction and do not serve its designed purpose. This is also due to poor construction practices, poor design procedures, and poor materials selection and handling. The Calabar-Itu highway in the southern part of Nigeria is a federal road (trunk A) that serves the states of Abia, Akwa Ibom, and Cross River states of the country and is of very high economic importance to the country. However, the road is in a state of despair and has been in its present dilapidated state for a very long time. Fig. 1 shows pictures of different failure conditions captured at different sections of the highway for the purpose of this research work. It shows that the highway collapsed due to multiple types of pavement failure, which included alligator cracking, longitudinal cracking, edge cracking, transverse cracking, potholes, and rutting. In pavement engineering, there are several layers that make the cross section of the roadway pavement, which include the subgrade layer (compacted insitu layer or in some cases borrowed and compacted), the subbase layer, the

Artificial Intelligence and Machine Learning in Smart City Planning
https://doi.org/10.1016/B978-0-323-99503-0.00020-X
Copyright © 2023 Elsevier Inc.

Fig. 1 Failure patterns on the studied pavement: (A) alligator cracking, (B) longitudinal cracking, (C) edge cracking, (D) transverse cracking, (E) potholes, and (F) rutting.

base course layer, and the surface layer (made of hot mixed asphalt). It is obvious that pavement failures are observed on the distressed surfacing course, but this does not mean that the failure originated from the topmost layer of the road pavement. The underlying layers may have suffered different degrees of distress which eventually affects the surface layer of asphalt. Meanwhile, the focus of the present research work is to examine the failed portions of the highway of case study, conduct materials sampling at the failed portion on the subgrade and subbase layers made of earth materials, conduct laboratory tests and generate multiple data on the mechanical properties of the pavement materials to be used in a smart and intelligent prediction of the CBR. With the above, this work intends to propose smart equations by making use of different AI-based techniques; GP, ANN, and EPR with which to design repairs and maintenance of the failed road and also to forecast its performance.

2. AI/ML in highway pavement subgrade and subbase construction and maintenance

There have been attempts made by researchers to apply AI/ML in modeling and forecasting infrastructural problems especially in the transportation environment. Transportation infrastructure is a complex component made up of different disciplines of interests with associated problems, for example, the pavement design, construction and performance, the traffic management and its associated indices of performance and sustenance, and maintenance of the infrastructure. However, all the components react in harmony to present a sustainable and durable infrastructural system. Meanwhile, in the present age of hi-tech and artificial intelligence, researchers in the field of civil engineering have absolved the benefits of AI/ML in solving the transport system problems in a smarter approach. Abduljabbar et al. [1] presented an overview on the application of AI in the transport system. This worked X-rayed the possibilities of applying AI-based learning techniques in solving the complex problems of the transport sector. Similarly, Tizghadam et al. [2] presented a work on ML in transportation exposing a data–driven solution that can deal with the conditions of the new environment. This work showed that ML has the ability to understand the hidden patterns of a historical data in order to model a transport system for a sustainable and smarter transport environment. Also, Simeunović et al. [3] had applied AI-based techniques in modeling the optimal number of public transport vehicles under mixed-traffic flow (MTF) conditions. MTF is common in the human society and it causes disruptions in schedules and travel time of the commuters in the flow of traffic. This literature studied and analyzed the traffic flow conditions on intersections to develop a smart transport model with high performance accuracy. Gangwani and Gangwani [4] also conducted a review work on the application of AI/ML in intelligent transport system, and in this work, various ML-based techniques applied in solving this problem were reviewed. Also, Iyer

[5] used AI-enabled techniques toward intelligent transportation. Iyer [5] studied a compilation of various issues troubling the transport industry classified under the intelligent transportation. In solving problems related to pavement foundation materials, Onyelowe and Shakeri [6] and Onyelowe et al. [7,8] attempted to apply AI-based model techniques to forecast the effect of adding waste-based ash materials as potential eco-friendly cementing materials on the behavior of the compacted subgrade. And very recently, multi-AI-based predictive models were developed for shrinkage limits of a treated problematic soil for compacted subgrade construction [9]. This produced high-efficiency performance models of over 90% with minimal error.

3. Application of AI/ML in subgrade and subbase CBR

The collected subgrade, subbase, and asphalt samples from the 26 failed portions (marking 52 tested samples for the CBR prediction and 26 samples for the MS prediction) of the pavement highway under study were subjected to preliminary experiments, which included PSD, Atterberg limits, compaction, and CBR tests in accordance with BS 1377-2 [10]. The soil samples were classified following the AASHTO guidelines presented in Table 1 and AASHTO numbers were assigned to every tested sample from the 26 points as presented in Table 2 for the purpose of the

Table 1 Statistical analysis of collected database.

	MDD (t/m^3)	OMC (%)	LL (%)	PL (%)	C (%)	AN	CBR (%)
Training set							
Max.	1.78	8.99	24.00	7.00	24.00	2.00	5.00
Min	2.00	14.86	49.00	15.00	44.46	7.00	33.00
Avg	1.86	10.11	32.62	10.31	31.09	3.08	14.96
SD	0.07	1.31	7.55	2.29	6.70	1.83	8.50
Var	0.04	0.13	0.23	0.22	0.22	0.60	0.57
Validation set							
Max.	1.91	9.14	27.00	9.00	23.30	2.00	20.00
Min	2.05	11.28	44.00	15.00	44.15	7.00	34.00
Avg	1.96	9.99	34.50	11.50	33.15	3.61	28.44
SD	0.03	0.59	6.88	2.19	7.41	2.29	4.96
Var	0.02	0.06	0.20	0.19	0.22	0.63	0.17

Table 2 Pearson correlation matrix.

	MDD	OMC	LL	PL	C	AN	CBR
MDD	1						
OMC	−0.11384	1					
LL	−0.18154	−0.02411	1				
PL	−0.07044	−0.03423	0.763187	1			
C	−0.29826	0.026813	0.773385	0.798563	1		
AN	−0.31198	−0.01541	0.827067	0.844649	0.892222	1	
CBR	0.880995	0.023697	−0.34442	−0.2551	−0.45719	−0.47206	1

intelligent prediction exercise. Asphalt extraction tests were conducted on 26 collected asphalt samples to determine the flow, the bitumen content, air void, and Marshall stability (MS) in line with the ASTM D6927-15 [11] and FMWH [12]. All the data collected from multiple exercises were recorded and tabulated.

4. Recent developments

4.1 Statistical analysis of the database

In a recent development on the application of AI in smart city planning and maintenance, the CBR of the subgrade and subbase failure situation of the flexible pavement under study was predicted and smart equations proposed for use in maintenance and design. A total of 52 soil samples (26 samples from the subgrade layer and 26 samples from the subbase layer) were tested to determine the following physical and mechanical proprieties: maximum dry density (MDD) t/m^3, optimum moisture content (OMC) %, liquid limit (LL) %, plastic limit (PL) %, fines' content (C) %, passing sieve #200, AASHTO number (AN), and California Bearing Ratio (CBR) %.

The measured records were divided into training set (34 records) and validation set (18 records). Tables 1 and 2 summarize the statistical characteristics and the Pearson correlation matrix of the database. It can be observed that MDD which has the lowest deviation in training and validation of the data also confirmed its closest correlation to the output parameter, CBR. This proves the relationship between density and the punching resistance ability expressed as CBR of both subgrade and subbase soils of pavement foundations. Finally, Fig. 2 shows the histograms for both inputs and outputs. The MDD also showed the best data distribution agreeable with the CBR distribution shown in Fig. 2.

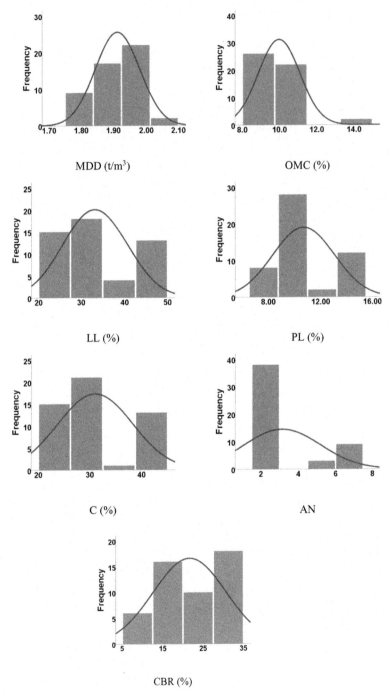

MDD (t/m³)

OMC (%)

LL (%)

PL (%)

C (%)

AN

CBR (%)

Fig. 2 Distribution histograms for inputs (in blue, gray in print version) and outputs (in green, light gray in print version).

4.2 Research program

Besides the traditional multilinear regression (MLR) technique shown in Eq. (1), three different artificial intelligent (AI) techniques were used to predict the CBR and MS of the tested soil and asphalt samples. These techniques are Genetic programming (GP), Artificial Neural Network (ANN), and Polynomial Linear Regression optimized using Genetic Algorithm which is known as evolutionary polynomial regression (EPR). All the three developed models were used to predict the values of California bearing ratio (CBR) using the measured treatment maximum dry density (MDD), optimum moisture content (OMC), liquid limit (LL), plastic limit (PL), fines' content (C), AASHTO number (AN), and Marshall stability (MS) using the measured bitumen content (BC) %, air voids (AV) %, and flow (F). Each model of the three developed models was based on different approaches (evolutionary approach for GP, mimicking biological neurons for ANN, and optimized mathematical regression technique for EPR). However, for all developed models, prediction accuracy was evaluated in terms of sum of squared errors (SSE).

$$C = \beta_0 + \beta_1 x_1 + \ldots + \beta_n x_n + \varepsilon \qquad (1)$$

where ε denotes the error of the model; $j = 0, 1, \ldots, n$ and β_j are the regression coefficients.

The following section discusses the results of each model. The validity and accuracies of the developed models were evaluated by comparing the (SSE) between predicted and calculated California bearing ratio (CBR) and Marshall Stability (MS) values.

4.3 Preliminary studies

The tested subgrade and subbase materials from 26 failed locations on the highway pavement showed a mixture of both poor and good properties even though they have failed their design usage. Six (6) out of the 26 locations showed poor material performance while 20 showed good materials performance. The implication of this outcome is that the pavement has failed due to poor materials characteristics. For instance, all the poorly rated locations had a CBR of less than 10% for the subgrade and less than 30% for the subbase, and are classified as A-7 group of soil in AASHTO classification method [13,14], which agrees with the standard specification for roads in Nigeria. Also, the MS results from the 26 locations on the failed highway pavement show that none of the locations showed MS below 3.5 kN, which is the MS standard specification for roads in Nigeria. The flow (F) and air void (AV) of 26 locations were within the standards; 2–4 mm and 3%–5%, respectively, specified for roads in Nigeria. Meanwhile, the bitumen content (BC) showed poor proportions of more than 90% beyond the standard 75%–82% BC specified for Nigerian roads and this posed a threat to the performance of the flexible pavement, hence its failed conditions. These indicate that the failure experienced on the flexible pavement resulted from poor materials constitution in the subgrade/subbase layers and BC overblown proportion of the pavement. Hence, the prediction of intelligent models to forecast best practices to manage the highway pavement for possible remedies and optimum performance.

4.4 GP prediction of California bearing ratio (CBR)

The developed GP model started with the one level of complexity and settled at five levels of complexity. The population size, survivor size, and number of generations were 100,000, 30,000, and 100, respectively. Eq. (2) presents the output formulas for (CBR), while Fig. 4A shows its fitness. The average error % of this equation is (16%), while the (R^2) value is (0.835).

$$CBR = 1.72 \, MDD^{4.9} - Ln\left(19500000 \, AN^2 + \left(\frac{2 \, LL}{3 \, OMC}\right)^{\frac{LL}{2}}\right) \quad (2)$$

4.5 ANN prediction of California bearing ratio (CBR)

A back propagation ANN with one hidden layer and (hyper tan) activation function was used to predict the same California bearing ratio (CBR) values. The used network layout is illustrated in Fig. 3. The average errors % of this model is (6.4%) and the corresponding (R^2) value is (0.976). The relation between calculated and predicted values is shown in Fig. 4B.

4.6 EPR prediction of California bearing ratio (CBR)

Finally, the developed EPR model was limited to hexagonal level, for 6 inputs; there are 924 possible terms $(462 + 252 + 126 + 56 + 21 + 6 + 1 = 924)$. GA technique was applied on these 924 terms to select the most effective 12 terms to predict the values of the California bearing ratio (CBR) values. The output is illustrated in Eq. (3) and its fitness is shown in Fig. 4C. The average error % and (R^2) values were (8.9%)–(0.951) for the total data sets, respectively.

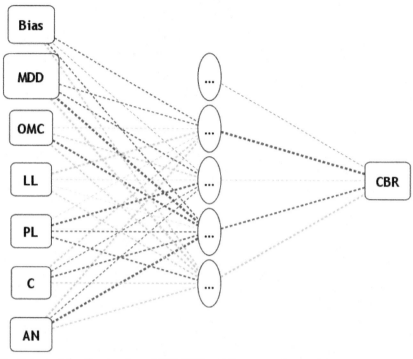

Fig. 3 Layout for the developed CBR ANN and its connection weights.

However, Table 3 shows the summary of the performance accuracies of the developed CBR models.

$$CBR = \frac{55.8\,AN^2 - 252\,AN}{LL} - \frac{28585\,MDD + 58420}{LL.AN}$$
$$+ \frac{321\,MDD^2}{AN} - \frac{29812}{LL.PL.AN} - \frac{PL.C.AN}{31.4} + \frac{PL^2.C}{39.3}$$
$$+ \frac{748\,AN}{C} + \frac{1030}{AN} - 17.6\,PL + 12.57 \tag{3}$$

5. Summary

This research presents three models using three (AI) techniques (GP, ANN, and EPR) to predict the California bearing ratio (CBR) using the measured treatment maximum dry density (MDD), optimum moisture content (OMC), liquid limit (LL), plastic limit (PL), fines' content (C), AASHTO Number (AN), and the Marshall stability (MS) using the

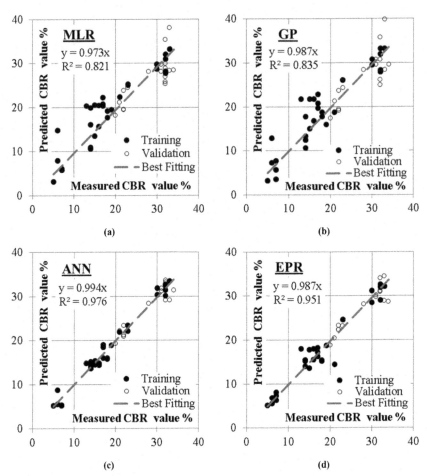

Fig. 4 Relation between predicted and calculated (CBR) values using the four developed models.

Table 3 Performance accuracies of developed CBR models.

Technique	Developed equation	Error %	R^2
MLR	Eq. (1)	16.1	0.821
GP	Eq. (2)	16.0	0.835
ANN	Fig. 3	6.4	0.975
EPR	Eq. (3)	8.9	0.951

measured bitumen content (BC) %, air voids (AV) %, and flow (F) from failed highway pavement requiring remedy management measures. The flexible pavement has been found failed due to disproportionate bitumen content (BC) and subgrade and subbase layer points that failed CBR requirements for road pavement foundation. The traditional multilinear regression (MLR) technique was used as a benchmark to evaluate the accuracies of the three (AI) techniques conducted on the multiple data collected from multiple test results on the failed locations of the pavement. The results of comparing the accuracies of the developed models could further be concluded in the following points:

For the California bearing ratio (CBR) predictive model:

- The prediction accuracies of the three (AI) techniques are more than or equal to the accuracy of the traditional MLR technique.
- The prediction accuracies of MLR and GP models are so close (83.9.0% and 84.0%) while the accuracies of ANN and EPR are close (93.6% and 91.1%) which gives an advantage to the EPR model because its output is a simple equation and could be applied either manually or implemented in software unlike the complicated output of the ANN which cannot be applied manually.
- Both Pearson correlation matrix and weights of ANN model showed that CBR value depends mainly on MDD, and other parameters are secondary and have almost equal weights.
- GA technique successfully reduced the 924 terms of conventional PLR hexagonal formula to only 12 terms without significant impact on its accuracy.
- Like any other regression technique, the generated formulas are valid within the considered range of parameter values; beyond this range, the prediction accuracy should be verified.

References

[1] R. Abduljabbar, H. Dia, S. Liyanage, S.A. Bagloee, Applications of artificial intelligence in transport: an overview, Sustainability 2019 (11) (2019) 189, https://doi.org/10.3390/su11010189.

[2] A. Tizghadam, H. Khazaei, M.H.Y. Moghaddam, Y. Hassan, Machine learning in transportation, J. Adv. Transp. 2019 (2019), 4359785. 3 pages, https://doi.org/10.1155/2019/4359785.

[3] M.M. Simeunović, V.Z. Bogdanović, P.M. Pitka, Z.M. Papić, D.M. Drašković, The model of the optimal number of public transport vehicles in mixed traffic flow conditions: a case study, Discr. Dyn. Nature Soc. 2021 (2021), 5276323. 19 pages, https://doi.org/10.1155/2021/5276323.

[4] D. Gangwani, P. Gangwani, Applications of machine learning and artificial intelligence in intelligent transportation system: a review, 2021, https://doi.org/10.1007/978-981-16-3067-5_16.

[5] L.S. Iyer, AI enabled applications towards intelligent transportation, Transp. Eng. 5 (2021), https://doi.org/10.1016/j.treng.2021.100083.

[6] K.C. Onyelowe, J. Shakeri, Intelligent prediction of coefficients of curvature and uniformity of hybrid cement modified unsaturated soil with NQF inclusion, Cleaner Eng. Technol. 4 (2021), https://doi.org/10.1016/j.clet.2021.100152.

[7] K.C. Onyelowe, A. Ebid, L. Nwobia, L. Dao-Phuc, Prediction and performance analysis of compression index of multiple-binder treated soil by genetic programming approach, Nanotechnol. Environ. Eng. (2021), https://doi.org/10.1007/s41204-021-00123-2.

[8] K.C. Onyelowe, M. Iqbal, F. Jalal, M. Onyia, I. Onuoha, Application of 3 algorithm ANN programming to predict the strength performance of hydrated-lime activated rice husk ash treated soil, Multiscale Multidiscip. Model. Exp. Des. (2021), https://doi.org/10.1007/s41939-021-00093-7.

[9] K.C. Onyelowe, A.M. Ebid, L.I. Nwobia, I.I. Obianyo, Shrinkage limit multi-AI-based predictive models for sustainable utilization of activated rice husk ash for treating expansive pavement subgrade, Transp. Infrastruct. Geotechnol. (2021), https://doi.org/10.1007/s40515-021-00199-y.

[10] BS 1377-2, Methods of Testing Soils for Civil Engineering Purposes, British Standard Institute, London, 1990.

[11] ASTM D6927-15, Standard Test Method for Marshall Stability and Flow of Asphalt Mixtures, ASTM International, West Conshohocken, PA, 2015, https://doi.org/10.1520/D6927-15.

[12] Federal Ministry of Works and Housing (FMWH), General Specification for Roads and Bridges, vol. II, Federal Highway Department, FMWH, Lagos, Nigeria, 1997, p. 317.

[13] AASHTO, Guide for mechanistic-empirical design of new and rehabilitated pavement structures [S], AASHTO, Washington, DC, 2004.

[14] AASHTO T 245, Standard Method of Test for Resistance to Plastic Flow of Asphalt Mixtures Using Marshall Apparatus, 2015.

Further reading

BS 5930, Methods of Soil Description, British Standard Institute, London, 2015.

CHAPTER ELEVEN

Machine learning algorithms-based solar power forecasting in smart cities

P. Tejaswi[a] and O.V. Gnana Swathika[b]
[a]School of Electrical Engineering, Vellore Institute of Technology, Chennai, India
[b]Centre for Smart Grid Technologies, School of Electrical Engineering, Vellore Institute of Technology, Chennai, India

1. Introduction

Solar energy is a promising and abundant resource in the world. It relies on various factors such as environment conditions, tropical region, and parameters like temperature, wind speed, irradiation, wind direction, and relative humidity. Efficient management and utilization of these resources are important for accomplishing the growing energy demand of the consumer. The limitation of solar energy is intermittent and unpredictable [1,2]. The solar photovoltaic systems act in standalone and in hybrid mode. Large storage system is required for standalone PV system; therefore, grid-connected photo voltaic systems are opted to get continuous power supply for a long time.

PV storage is a part of the modern grid-connected system. In the grid-connected PV system, there is a mismatch between generation of power and consumption of power which leads to causing fluctuations in the entire power system. The accurate prediction of solar energy is very important to overcome the above problem. Depending on the PV energy availability, loads can be scheduled with predefined priorities and storage actions are taken accordingly. For this, the accurate forecasting of PV output depends on reliable prediction models. The prediction of solar energy also depends on levels of solar irradiation and climatic conditions of that region in [3]. Hence there is a need to schedule the load operation and forecast PV output efficiently at the consumer. Various machine learning algorithms are considered for predicting the solar irradiance based on the parameters used, length of the data set, and usage details. The main benefits of accurate solar energy forecasting are dispatch-ability, low cost, and efficiency for energy consumers and utilities. The secure and reliable operation of power system depends on data of the load consumption and planning of renewable generation forecast. The accurate prediction of solar energy aids energy sector to reduce fluctuations in power and maintain reliability of the overall power system. The health of the power system is preserved by monitoring the forecasted information continuously [4].

2. Overview of machine learning

Human beings are the smart and advanced species on this planet because they can think, evaluate, and solve complicated issues. On the other hand, artificial intelligence is in the initial stage when compared to human intelligence in many aspects. Then the purpose of machine learning is to take decisions based on the data available with efficiency. Research has been going on in technologies like machine learning, artificial intelligence, and deep learning to solve real-world complex issues. The decisions are taken by machines based on the data to automate the process. Some real-world problems use these data-driven decisions, where programing logic cannot be used directly. That is why there is a need for machine learning to solve real-world issues with efficacy at a large scale.

Machine learning is a part of artificial intelligence which helps the computer systems to sense the data and take proper decision for forecasting. Machine learning extracts patterns from raw data by using algorithms. Machine learning allows computer systems to learn through experience rather than explicitly programmed. Machine learning models consist of learning algorithms which executes some task and enhance their performance over time with experience.

Machine learning is the fastly expanding technology in the present world. Some researchers named that we are in the golden era of artificial intelligence and machine learning. Real-world complex problems are solved by the machine learning algorithms, which are not resolved with the help of conventional methods in Obulesu et al. [5]. The real-world applications of machine learning algorithms are prediction of weather, emotion analysis, detection and prevention of error, sentiment analysis, recognition of object, stock market forecasting, speech synthesis and recognition, customer segmentation, smart city planning, fraud detection and prevention [6].

3. Methodology

The various steps involved in the machine learning model are data collection and preprocessing, dividing data for training and testing, building model, training model, testing the model, and calculation of performance metrics as shown in Fig. 1.

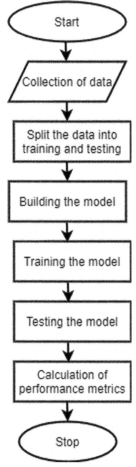

Fig. 1 Basic prediction model. *(No permission required.)*

3.1 Data collection and data preprocessing for accurate energy prediction

Appropriate data set collection is the primary step of designing a forecasting model. The historical information of solar energy is obtained from Kaggle website and the data utilized for feeding the machine learning algorithm are day of year, year, month, day, first hour of the day, temperature, wind speed, visibility, and relative humidity.

3.1.1 Data preprocessing

The important task after collecting the data is preprocessing the data. It will transform the selected data into a form which is fed to machine learning algorithms. Preprocessing the data improves prediction accuracy of machine learning algorithm [7].

Data preprocessing techniques
 (i) *Scaling*:
   The data with attributes consisting of varying scale cannot be fed to machine learning algorithm; hence the data need to be rescaled. Data rescaling ensures that attributes are at same scale. The attributes in the data set are rescaled in between 0 and 1.
(ii) *Standardization*:
   This method is used to convert the attributes of the data with a Gaussian distribution. For a standard Gaussian distribution, this method will have a mean of 0 and an SD of 1. This method is applied to logistic regression and linear regression. If the input attribute has Gaussian distribution, then the machine learning algorithm yields good results with the rescaled data.

3.1.2 Building the model

The various parameters are tuned to design an efficient prediction model. The forecasting algorithms are trained and tested based on the prediction accuracy and performance metrics.

3.1.3 Training and testing the model

The entire data set is divided into training and testing section during the training phase. Generally for evaluating the prediction models, eighty percent of the data set is used for training and twenty percent of the data set is used for testing.

3.1.4 Machine learning models

(i) *Random forest regression algorithm:*

It is a supervised machine learning algorithm which contains decision trees in various subsets of the given data set. The prediction accuracy of that data set is improved by taking the average of decision trees. The random forest collects the prediction from each tree and implements the decision by considering the majority votes of predictions. The final prediction result is the most voted prediction result. The more the number of trees in the forest the higher is the accuracy and also prevents the overfitting problem [8].

(ii) *Artificial neural network (ANN):*

ANN is a forecasting algorithm utilized for forecasting of solar energy. The ANN model considers the past data with various parameters such as temperature, month of the day, wind speed, sky cover, relative humidity, and year. Further, the data set is classified into training set and testing. The parameters of ANN have various combinations of number of hidden layers, neurons, kernels, and activation functions to obtain the best optimal value. After a number of trials, the trained model with least error is selected for forecasting [9,10].

Performance metrics:

Various performance metrics are used to measure the accuracy of solar power forecasting. Standard performance metrics aids in evaluating forecasting model. Mean-absolute error, mean-square error, mean-absolute percentage error, and root mean-square error are the mostly used performance metrics. These performance measures have their own emphasis and features. Prediction evaluation is accomplished based on the choice of appropriate performance metrics according to the special conditions [11,12].

Mean-absolute error (MAE)

It is difference between the actual and the predicted value of data. It is expressed in Eq. (1)

$$\mathrm{MAE} = \frac{1}{N} \sum_{k=1}^{N} |P - A| \tag{1}$$

where N is the samples number, P is the predicted data, and A is the actual data.

Mean-absolute percentage error (MAPE)

Mean absolute error is expressed in percentage. It is expressed in Eq. (2)

$$\text{MAPE} = \frac{1}{N} \sum_{k=1}^{N} \frac{|P - A|}{PN} \tag{2}$$

where N is the samples number, P is the predicted data, A is the actual data, and PN is the nominal value.

Root mean-square error (RMSE)

It is expressed in Eq. (3)

$$\text{RMSE} = \sqrt{\frac{1}{N} \sum_{k=1}^{N} \left(\frac{P - A}{P}\right)^2} \tag{3}$$

where N is the number of samples, P is the predicted data, and A is the actual data.

4. Results and analysis

Random forest regression algorithm is built with number of estimators 600, number of features used are 4. The total data set is categorized into eighty percent data as training and twenty percent data as testing. The input variables in the data set are day, month, first hour of the year, year, temperature, wind speed, sky cover, visibility, and relative humidity. The target variable in this model is solar power in watts. Data preprocessing includes data cleaning and standardization. Data cleaning fills the missing values of the data set with mean of that particular column of the data set. The scaling technique standardization is used to obtain standard deviation of one and mean of zero. Accuracy metrics is calculated for the test data in this model which is tabulated in Table 1. Fig. 2 depicts the training predictions versus

Table 1 Comparison table for the prediction models showing performance metrics

Algorithm	Correlation coefficient	Mean-absolute error	Mean-square error	Root mean-square error
Random forest regression	0.91	12300.62	10635.71	0.2968
Artificial neural network	0.899	2257.76	3476.2314	0.2093

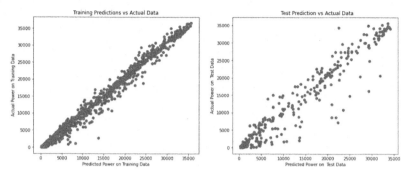

Fig. 2 Training prediction versus actual data. *(No permission required.)*

actual data and test prediction versus actual data. All the points in this model lie near to $x=y$ line which tells the model is having good accuracy for predictions.

Artificial neural network algorithm is built with epochs 180 and verbose is 2. Relu is the activation function used in ANN model. The total data set is classified into 75 percent data as training and 25 percent data as testing. The input variables in the data set are day, month, first hour of the year, year, distance from solar noon, temperature, wind speed, sky cover, visibility, and relative humidity. The target variable in this model is solar power in watts. Data preprocessing includes data cleaning and standardization. Data cleaning fills the missing values of the data set with mean of that particular column of the data set. The scaling technique standardization is used to obtain standard deviation of one and mean of zero. Performance metrics is calculated for the test data in this model which is tabulated in Table 1. Fig. 3 depicts the training predictions versus actual data and test prediction

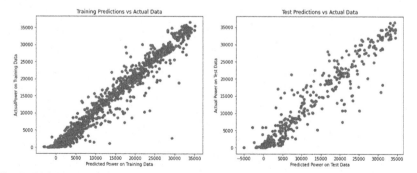

Fig. 3 Training prediction versus actual data. *(No permission required.)*

versus actual data. All the points in this model lie near to x = y line which tells the model is having good accuracy for predictions.

In Table 1, the correlation coefficient for the random forest regression and artificial neural network models is around 0.9, which shows that the accuracy of the prediction models is good. The comparison of mean-absolute error, mean-square error, and root mean-square error is made for the random forest regression and artificial neural network models.

5. Conclusion

The optimum utilization of renewable source is assessed by prediction of solar power accurately. The machine learning models such as random forest algorithm and ANN techniques are assessed to determine the prediction models accurately using python. Comparison of these forecasting algorithms is carried out in terms of correlation coefficient, mean-absolute error, mean-square error, and root mean-square error. The random forest and artificial neural network machine learning models give the accurate prediction of solar energy which is important in these days where there is a requirement for PV storage as part of the modern grid connected system. To enhance the abilities of prediction, a new optimization algorithm can be used which can combine with the proposed forecast technique. Thus, the forecast of biomass energy, tidal energy, geothermal energy, wave energy, and hydraulic power could be prospective fields for the forthcoming work.

References

[1] E. Craparo, M. Karatas, D.I. Singham, A robust optimization approach to hybrid microgrid operation using ensemble weather forecasts, Appl. Energy 201 (2017) 135–147, https://doi.org/10.1016/j.apenergy.2017.05.068.
[2] S. Sperati, S. Alessandrini, P. Pinson, G. Kariniotakis, The Weather intelligence for renewable energies benchmarking exercise on short-term forecasting of wind and solar power generation, Energies 8 (9) (2015) 9594–9619, https://doi.org/10.3390/en8099594.
[3] A. Agüera-Pérez, J.C. Palomares-Salas, J.J. González de la Rosa, O. Florencias-Oliveros, Weather forecasts for microgrid energy management: Review, discussion and recommendations, Appl. Energy 228 (2018) 265–278, https://doi.org/10.1016/j.apenergy.2018.06.087.
[4] L. Gigoni, A. Betti, E. Crisostomi, A. Franco, M. Tucci, F. Bizzarri, D. Mucci, Day-ahead hourly forecasting of power generation from photovoltaic plants, IEEE Trans. Sustain. Energy 9 (2) (2018) 831–842, https://doi.org/10.1109/TSTE.2017.2762435.
[5] M. Obulesu, M. Mahendra, ThrilokReddy, A review of studies on machine learning techniques, in: International Conference on Inventive Research in Computing Applications (ICIRCA), 2018.

[6] H. Wang, Y. Liu, B. Zhou, C. Li, G. Cao, N. Voropai, E. Barakhtenko, Taxonomy research of artificial intelligence for deterministic solar power forecasting, Energ. Conver. Manage. 214 (2020), https://doi.org/10.1016/j.enconman.2020.112909.

[7] M.Q. Raza, M. Nadarajah, C. Ekanayake, On recent advances in PV output power forecast, Solar Energy 136 (2016) 125–144, https://doi.org/10.1016/j.solener.2016.06.073.

[8] M.W. Ahmed, M. Mourshed, Y. Rezgui, Tree based ensemble methods for predicting PV generation and their comparison with support vector regression, Energy 164 (2018) 465–474.

[9] P. Pawar, M. TarunKumar, K. Panduranga Vittal, An IoT based Intelligent Smart Energy Management System with accurate forecasting and load strategy for renewable generation, Measurement 152 (2020) 107187.

[10] C. Wan, J. Zhao, Y. Song, Z. Xu, J. Lin, Z. Hu, Photovoltaic and solar power forecasting for smart grid energy management, CSEE J. Power Energy Syst. (2015) 38–46, https://doi.org/10.17775/CSEEJPES.2015.00046.

[11] R. Ahmed, V. Sreeram, Y. Mishra, M.D. Arif, A review and evaluation of the state-of-the-art in PV solar power forecasting: Techniques and optimization, Renew. Sustain. Energy Rev. 124 (2020), https://doi.org/10.1016/j.rser.2020.109792.

[12] M.K. Behera, I. Majumder, N. Nayak, Solar photovoltaic power forecasting using optimized modified extreme learning machine technique, Eng. Sci. Technol., Int. J. 21 (3) (2018) 428–438, https://doi.org/10.1016/j.jestch.2018.04.013.

CHAPTER TWELVE

Smart grid: Solid-state transformer and load forecasting techniques using artificial intelligence

Dharmendra Yadeo[a], Sachidananda Sen[a], and Vigya Saxena[b]
[a]Department of Electrical and Electronics Engineering, SR University, Warangal, Telangana, India
[b]Indian Institute of Technology IIT (ISM), Dhanbad, Jharkhand, India

1. Introduction

Smart grid is a key technology in realization of the vision of smart city. It is an electricity network which can intelligently combine actions of all the users connected to it, in order to efficiently deliver sustainable, economic, and secured electricity supplies. It consists of various smart devices through which energy demand and supply can be managed efficiently. It enables integration of various renewable energy sources (RES), distributed generation, and energy storage devices feasible which will reduce the environmental impact due to reducing dependency on conventional power plant (thermal power plant). With intelligent operation of various assets in the smart grid, their performance can be optimized for better delivery system depending on consumer needs. It will also improve reliability and resiliency of the system along with improved power quality of the electricity network.

Solid-state transformer (SST) can play very crucial role in realization of smart grid. SST can be equipped with features such as high-power density and provision to easily interface renewable energy sources to the DC bus. Due to availability of DC port, charging station for electric vehicle can be easily set up. Moreover, AC grid can also be connected with SST, which will serve many existing AC loads. For improving reliability in the power generation system, battery equipped with bidirectional converter as part of SST can be used efficiently. The bidirectional converter can also be used to facilitate the power transfer between DC bus and AC buses of a distribution network or microgrid.

In order to deliver power at minimum price to the customer, with good reliability, safely and continuously, load forecasting is very essential. Load forecast can be of different durations, viz., long term (more than one year), medium term (one week to one year), short term (one hour to one week), and very short term (one minute to one hour). Various methods have been given in literature, which can be broadly categorized into two categories: statistical approach and artificial intelligence (AI)-based approach. Statistical method involves mathematical combination of previous loads and current or previous values exogenous factor such as weather. The methods' results are satisfactory when load is behaving linearly, but when their operation becomes nonlinear, it gives unsatisfactory results. On the other hand, AI-based approach can handle complexity and nonlinearity well. Among various AI-based techniques, algorithm obtained using artificial neural network (ANN)-based approaches are quiet interesting and satisfactory.

2. Power distribution system

In the existing AC system, electric power basically comprises three stages: generation, transmission, and distribution. Fig. 1 shows the schematics of an existing AC power system. Conventionally, power is generated remotely at low voltage in a power plant. Then with the aid of a line frequency transformer, which operates at grid frequency of 50 Hz, it is transmitted at different voltage levels to various loads.

Even though conventional line frequency transformers have become the backbone of the existing power system, they suffer from various limitations as mentioned below:

- They have huge size and weight. Transformer oil when exposed to environment can be hazardous.
- Core saturation due to an overloading may produce large inrush current and distorted secondary side voltage.
- Losses in the transformer will be increased due to harmonics in the output current waveform, which can be due to various power electronics loads connected at the secondary side.
- Conventional transformers are normally designed to operate at full load. Their efficiency will decrease if operated at 30% load, which is the average loading condition in a distribution environment.
- They do not perform any other function besides isolation and voltage conversion.

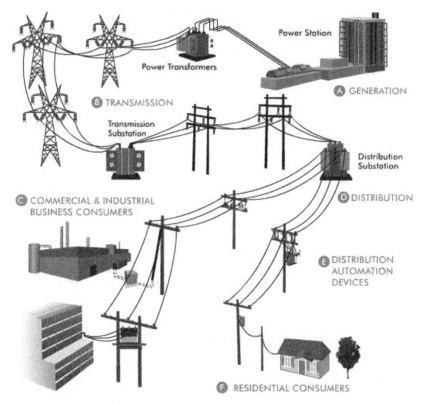

Fig. 1 Power distribution system.

With the increased penetration of renewable energy sources (RESs), existing grid functionality has also been challenged. A large number of distributed energy resources (DERs) in the existing AC infrastructure have caused increase in voltage in low-voltage feeders. In order to deal with various issues related with power quality like voltage sag, voltage swell, waveform distortion, transients and voltage imbalances, many devices such as volt-amp reactor, static synchronous compensators (STATCOM), capacitor bank, dynamic voltage restorer (DVR), unified power flow controller (UPFC), unified power quality conditioner (UPQC) are used externally [1–3].

For dealing with the various issues associated with the existing AC system, the role of solid-state transformer (SST) becomes more prominent. It can not only serve as a replacement of the conventional line frequency transformer but can also resolve almost all the issues related with the power quality in the existing AC system.

2.1 Future power distribution system in smart city

Nowadays, urban population is increasing with greater pace due to migration of people from villages to city. It imposes various environmental, economic, and societal challenges on the city administration. With various technological advancements, effort is to make the city a better place to live by overcoming all the challenges it may face in the near future. Uninterrupted and sustainable power supply is the need of the hour. As the trend toward realizing the vision of smart city is growing day by day, it becomes necessary to equip smart city with smart power distribution network. Smart grid and microgrid if intelligently controlled will play a crucial role in excellent operation of power network. In Sen and Kumar [4,5], various control techniques have been discussed for the efficient operation of microgrid.

Generalized block diagram of a microgrid (MG) which comprises renewable energy sources, conventional AC grid, two DC buses, various DC loads, and energy storage system is shown in Fig. 2. Medium voltage DC (MVDC) bus is connected to various RESs such as solar and wind. As RESs are intermittent in nature, to improve the reliability, microgrid is equipped with battery and super capacitor for energy storage. Battery has slow dynamic response, so they will provide power during power

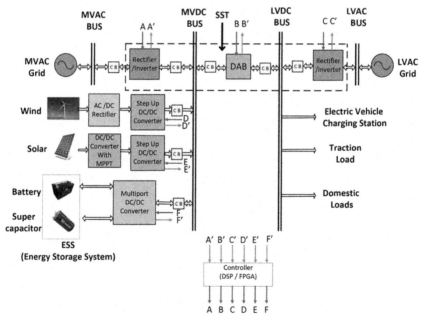

Fig. 2 Future power distribution system.

outages. Whereas, super capacitors have fast dynamic response, thus, they will operate during transient stages, such as voltage sag and swell. Thus, combination of both battery and super capacitor, having the features of high energy density and high power density, respectively, can be utilized to improve the reliability of the microgrid [6–11].

On low-voltage DC (LVDC), various DC loads are connected. Conventional medium-voltage AC grid and low-voltage AC grid can be integrated with the aid of solid-state transformer. SST comprises two bidirectional converter (rectifier or inverter) connected at each end of the SST and dual active bridge, which serves as a mediator between two converters and DC buses operating at same or different voltage levels. SST can be used for integrating the two (AC and DC) grids. Dual active bridge (DAB) which acts as mediator between two converters in SST is a DC-DC converter. It comprises high-frequency transformer (HFT) which is used for isolation. In Yadeo et al. [12–14], many DC to DC converters are given, having their own advantages and limitations. Among all the converters, dual-active bridge which consists of two H-bridges on both the sides of HFT is the most popular, due to its simple circuit. Nowadays, research on multilevel dual-active bridge converters are also increasing. Multilevel converters can be used for high-power transmission and can sustain high voltages. For better control aspect, each converter can be controlled by controllers having signal processing capabilities of FPGA or DSP. It will sense input and output data from the converter which will enable efficient operation of the converter.

2.1.1 Solid-state transformer

In recent decades, due to advancements in semiconductor switching device technology, solid-state transformer or power electronic transformer (PET) has gained much attention. Fig. 3 shows the block diagram of solid-state transformer. It is basically a power electronics device operating at higher frequency and utilizes high-frequency transformer for providing isolation. It

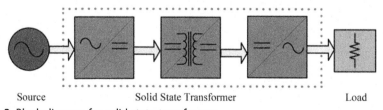

Source   Solid State Transformer   Load

Fig. 3 Block diagram for solid-state transformer.

replaces the conventional line frequency transformer, and if controlled wisely, it also provides many additional features which are not available in conventional line frequency transformer [15–23].

With advancement of development in technology for power conversion at high-voltage level, emphasis is more on choosing right switching devices for the realization of SST. In the last 50 years, trend was growing toward Si MOSFET and Si IGBT for high-frequency power conversion. However, in these devices, switching frequency cannot be increased beyond some threshold limit due to rising switching losses, which poses major challenge in increasing power density and compactness of converter used in SST implementation. Nowadays, research and development in wide-band gap (WBG) devices such as Si carbide (SiC) and gallium nitride (GaN) materials are gaining momentum. WBG-based devices have very less turn *ON* and turn *OFF* time. It reduces their switching losses to a larger extent, compared to Si-based devices. Thus, with WBG devices, power density in SST can be increased, leading to compactness in the converter's size.

2.1.2 Various enhanced features of SST

- **High-power density**: As SST is composed of high-frequency transformer, it effectively reduces the size of transformer with increase in operating frequency for rated power.
- **Instantaneous voltage regulation**: Grid problem such as voltage sag and swell can be resolved with enhanced controlling features without the aid of any extra major components.
- **DC output ports**: With the provision of DC ports, various DC loads can be connected.
- **RES integration**: Availability of DC link and facilitate integration of various RES.
- **Grid integration**: Easy interface with smart grid network is feasible as AC voltage is converted to DC voltages. Thus any issue regarding frequency mismatching can be avoided.
- **Bidirectional active and reactive power flow control**: As converters in SST are bidirectional, it enables bidirectional power flow control.

2.1.3 Different Configurations of SST

SSTs are broadly classified into four categories: A type, B type, C type, and D type depending on their intermediate stages. Fig. 4 shows different configurations of SST.

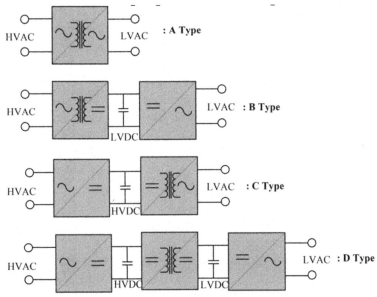

Fig. 4 Different configurations for solid-state transformer.

A-type SST involves a single-stage conversion where there is direct conversion from high-voltage AC to low-voltage AC with isolation transformer. There is absence of DC link voltage in A-type, which makes voltage regulation and solar integration difficult. Moreover, to mitigate the harmonics, filter requirement would also be large.

There are two stages (AC to DC and DC to AC) in B- and C-type configurations. In B type, lowvoltage DC link is present and isolation is provided in the AC to DC stage to separate the high-voltage side from the low-voltage side. Whereas in C type, a high-voltage DC link is present and the DC to AC stage provides isolation.

D-type configuration involves three-stage conversion, comprising AC to DC, DC to DC, and AC to AC stages, where isolation is provided by the DC to DC stage. It combines the advantages of both B-type and C type configurations. In the future smart grid, features such as current limiting, voltage regulation, maintaining power quality, connecting energy storage, reactive power support for grid, and interfacing distributed resources can be realized using the D-type configuration.

3. Load forecasting

In today's scenario, electricity can be bought and sold like any other commodities. It involves various trade agreements between the firms involved in generation, transmission, and distribution of power. Any abrupt jump in electricity load will increase its price. For firms which are not prepared for dealing with price escalation may have to bear heavy loss. The reason behind the increase in price is that it cannot be stored in bulk and storing excess power in energy storage elements will increase cost on the system. Also, most of the power equipment are designed to operate efficiently at the rated load. Under loading or overloading of the power equipment will lead to its inefficient operation, which may cause heavy burden on the power plant. So, it becomes necessary to know the load in advance which will help to optimize operation of the power equipment, scheduling power generation, improving reliability, and efficiency of the system. For making a good forecast, model analysis of the load data is desired. Fig. 5 shows monthly load demand where it can be observed that there is a repetition pattern in the weekly load demand.

There are many factors which affect the loads. It may change as per the season, time, festivals, and growth of area. Load forecasting plays a very crucial role in the nonpower sector or different industries for planning their daily operation. It will enhance the overall functioning of smart grid and smart city. Load scheduling, infrastructure, and maintenance can be planned well using efficient load forecasting techniques. Power quality of the power

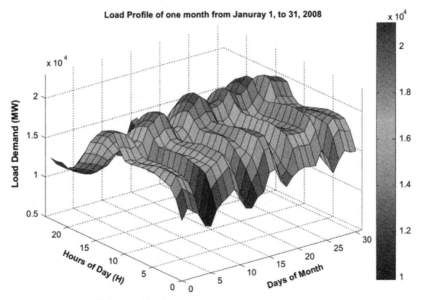

Fig. 5 Monthly load demand [24].

system deteriorated with the phenomenon such as voltage sag, swell, spikes, and outages which may happen due to improper loading. With intelligent load-forecasting techniques, power quality of power system can be maintained properly. Now, with the incorporation of distributed energy resources, utilities can design efficient methods for power flow regulation using advance load forecasting techniques.

Load forecasting methods are broadly classified into two categories:
1. Statistical approach
2. Artificial intelligence-based approach

3.1 Statistical approach

The statistical methods take into account previous load and exogenous factors. These techniques are very attractive as they have some physical interpretation which enables system operators or engineers to analyze the load behavior. However, these techniques suffered due to their least ability to model loads which are nonlinear.

In statistical methods, loads are represented as mathematical model of various factors. Models may be additive or multiplicative. Additive model may be represented as

$$L_t = L_t^b + L_t^w + L_t^s + \varepsilon_t$$

where L_t^b = base load, L_t^w = weather-dependent part of load, L_t^s = special event-dependent load, ε_t = noise.

There are various statistical methods, for example, similar day method, exponential smoothing, regression methods, autoregressive model, autoregressive moving average (ARMA) model, autoregressive integrated moving average (ARIMA) model, and time series model with exogenous variable. Some of these methods are:

- **Similar day method:** In this method, historic data of a given day having similar characteristics are taken into account. This method is not carefully calibrated so it gives error in forecasting.
- **Exponential smoothing:** In this method, weighted sum of previous observations is used. These can be classified as single, double, or triple exponential smoothing depending on the number of parameters it takes into account. Single-exponential smoothing has single parameter called as alpha which is also known as smoothing factor. In double-exponential smoothing, in addition to "alpha," another parameter named "beta" is taken into account that controls the trend. Trend may be additive or

multiplicative. In triple-exponential smoothing, third parameter is also taken into account which is known as "gamma."

- **Regression method:** Linear regression method uses two or more variables to find relation between them. After knowing the relation between different variables, it is assumed that relation between parameters remains the same throughout analysis. Different variables may be load and other factors which affect it, for example, weather, customer class, and day type. It utilizes curve fitting tool to find different values of coefficients. Performance of algorithm is evaluated by measuring several parameters such as Least Squared Error (LSE), Mean Squared Error (MSE), Root Mean Squared Error (RMSE), Mean Absolute Percentage Error (MAPE), and Mean Absolute Percentage Deviation (MAPD).

3.2 Artificial intelligence-based technique

In artificial intelligence technique, artificial neural network (ANN) is most widely used in load forecasting. With ANN, input and output mapping can be performed without any complex computation. Depending on the pattern reorganization, output behavior can be learnt. ANN performance under uncertain input values is very robust. It has fast convergence speed, low computational complexity, and less training period. ANN consists of three layers, namely, input layer, hidden layer, and output layer as shown in Fig. 6. Each layer is connected with intermediate layer using neurons which are also called as processing elements. All neurons have synaptic weights that are adjusted for mapping the input–output relationship. Weighted inputs are applied to activation function before obtaining output function.

Output can be calculated as

$$A_i = g\left(\sum_{i=0}^{n} W_{ji} {}^* a_j \right)$$

where A_i is the output of network, W_{ji} is the weight of jth neuron in ith layer, and a_j is the input of neuron.

Activation function decides how weighted input is converted to output depending on the type of function use in the activation process. It can be a logistic function, step function, linear function, sigmoid function, or tangent hyperbolic function. Activation function is a two-step process which involves sum of weighted input and transfer function. Transfer function depends on the type of problem. It can be unipolar step function, bipolar

Architecture for Neural Network

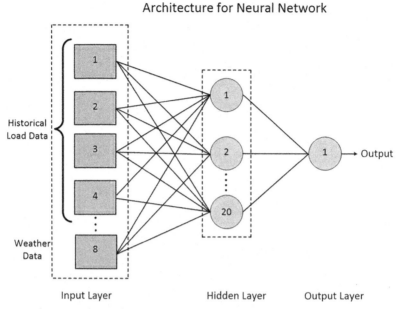

Input Layer   Hidden Layer   Output Layer

Fig. 6 Architecture for artificial neural network.

step function, unipolar linear function, bipolar linear function, unipolar sigmoid, bipolar sigmoid, or Gaussian radial basis.

ANN can have multiple-layer perceptron or single-layer perceptron network. Multilayer perceptron neural network has the ability to learn complex relationship which has more than one hidden layer in between input and output layer. Depending on the network architecture, ANN can be divided into feed forward neural network or feedback neural network. Feed forward neural network may have one or more layers between input and output as shown in Fig. 7. It is the simplest form of network, and in this network, flow of information is from input to output through the hidden layer without any loop.

In feedback neural network, close loop enables bidirectional flow of information between input and output. Fig. 8 shows feedback neural network. In dynamic and complex processes, feedback neural networks are highly useful.

Among AI-based algorithms, artificial neural network has relatively good performance and its implementation is also straight-forward.

Classification of ANN can be done on the basis of their architecture, processing, and training. Neural connection is described by the architecture

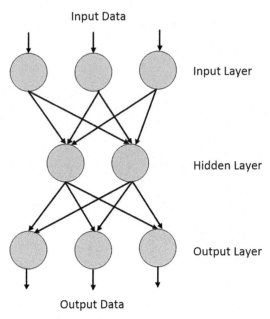

Fig. 7 Feed forward neural network.

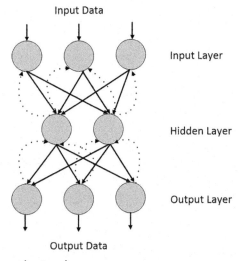

Fig. 8 Feedback neural network.

of any NN. Input and output of NN is determined by the number of connected layers between them. Processing of any NN is described by its relation between the output and their corresponding input and weight. In order to train the neurons in a neural network, some learning techniques are required. This learning can be supervised or unsupervised. Back propagation is a well-known architecture for short-term load forecasting which takes into account supervised learning. In supervised learning, error between the target value and output value is reduced by updating the weight value. It reduces the mean square error. In unsupervised learning, there is no target value and the neuron weight is adjusted by the system itself according to different inputs.

ANN can be broadly classified into two groups depending upon the number of output nodes. First group has only one output node, whereas the second group has many output nodes. The first group will forecast only one parameter, e.g., next hour load or next day peak load, whereas for the second group, they may determine whole day load profile depending on the number of output profiles they have, e.g., 24 output nodes will give 24 h load profile.

In order to improve the performance of load forecasting, many hybrid load forecasting techniques are used. These take into account two or more algorithms by taking advantage of each algorithm. In literature, many algorithms based on neural network are integrated with other optimization techniques to enhance the performance. Factors such as learning algorithm, network structure, network parameter, and historical load data quality decide the forecast model accuracy. For a short-term load forecasting, hybrid techniques along with ANN are fuzzy logic, genetic algorithm, regression technique, expert system, wavelet and time series, support vector machine, artificial immune system, gradient-based learning techniques are used.

Few methods which are based on optimization algorithm and other artificial intelligence techniques are as follows:

(1) Fuzzy logic and genetic algorithm:

In fuzzy logic, many logic values are used as input which ranges between 0 and 1. Fuzzy models have various capabilities such as manipulating, recognizing, interpreting, and representing the data that lack certainty and are vague. Implementation of fuzzy logic involves the process of fuzzification which processes on various inputs whereas defuzzification involves extraction of precise output which can be interpreted by human being. In [25,26] fuzzy logic is used to predict short-term load by considering the weather sensitive data and historical load data. Fuzzy logic does not require complex mathematical function for mapping of the input and output.

Genetic algorithm is a search-based optimization technique which works on the principle of genetics. Optimized solutions are obtained for any problem which may take otherwise too much time to solve.

In Ray et al. [27], genetic algorithm with back propagation has been used along with the artificial neural network for load forecasting.

(2) Regression technique and expert system:

When qualities such as reasoning, explaining, and knowledge enhancement according to new information are embedded into any system, then it becomes an expert system. Expert system works on the rules and procedures feed by human experts into software. It incorporates various if/then statements which are coded into the software platform to formulate any decision. In Mori and Kosemura [28], hybrid load forecast model using global optimal regression tree and multilayer perceptron is used. It has better convergence rate and also accelerates the training of neural network.

(3) Support vector machine (SVM) and artificial immune system:

It is a regression and classification tool which is used for performing nonlinear mapping of various data into multidimensional space. They have flexible structures and do not rely much on heuristics. In Chen et al. [29], SVM is used for load forecasting.

(4) Self-organizing map (SOM):

It is a type of neural network where neurons are trained using unsupervised learning and during the training and weight of neurons is adjusted by competitive learning which reduces the dimension and data. It is a three-stage process which involves competition, cooperation, and adaption. In Martín-Merino and Román [30], an algorithm to split the time series using self-organizing map is proposed. In this study, a new model which takes into account the SOM neighborhood relation and builds an input space partition for predicting a target and avoiding overfitting is proposed. With this method, maximum electricity demand can also be predicted.

(5) Extreme learning machine:

Extreme learning machine is a type of feed forward neural network where weight of the nodes is not tuned. The hidden nodes are randomly assigned. Models based on extreme learning are faster than the networks which use back propagation. These are used for classification, clustering, regression, and sparse approximation. In Dash and Patel [31], extreme learning machine method is used for forecasting the day ahead of load.

(6) Convolution neural network (CNN):

CNN is a deep learning algorithm which processes over image to extract and differentiate information. There are many layers (input layer, convolution and pooling layer, flattening layer, fully connected layer, and output layer) existing in a CNN. With CNN, two dimensional maps are converted into one-dimensional array. Recently, in Tudose et al. [32], CNN has been proposed to forecast load which also takes into account the socioeconomic impact of COVID-19 along with other exogenous factors such as historical data and weather change.

4. Summary

Solid-state transformer and load-forecasting techniques are going to be key technology in implementation of the vision of smart grid. SST will improve the power density and is equipped with many additional features, which is not available in the conventional transformer. SST will allow to integrate various renewable energy sources easily to DC bus and due to AC and DC stages, as well as both types of loads (AC and DC loads) can be connected. In the near future, many distributed energy resources will be connected with grid resulting into formation of microgrids and active distribution networks. The number of loads will also increase, and with the rising competitiveness in electricity market, it becomes necessary to predict load accurately. In this chapter, various functionings of solid-state transformer along with the load-forecasting methods that are based on the statistical and artificial intelligence-based approach have been discussed. SST when combined with the enhanced load forecasting techniques can bring efficient operation of smart grid that will make cities smarter.

References

[1] G.F. Reed, B.M. Grainger, A.R. Sparacino, Z. Mao, Ship to grid: medium—voltage DC concepts in theory and practice, IEEE Power Energy Mag. 10 (6) (2012) 70–79.
[2] N.G. Hingorani, High-voltage DC transmission: a power electronics workhorse, IEEE Spectrum 33 (4) (1996) 63–72.
[3] G.L. Kusic, G.F. Reed, J. Svensson, Z. Wang, A case for medium voltage DC for distribution circuit applications, in: 2011 IEEE/PES Power Systems Conference and Exposition, Phoenix, AZ, 2011, pp. 1–7.
[4] S. Sen, V. Kumar, Microgrid modelling: a comprehensive survey, Ann. Rev. Contr. 46 (2018) 216–250.
[5] S. Sen, V. Kumar, Microgrid control: a comprehensive survey, Ann. Rev. Contr. 45 (2018) 118–151.
[6] S.K. Kollimalla, M.K. Mishra, N. Lakshmi, A new control strategy for interfacing battery supercapacitor storage systems for PV system, in: IEEE Students' Conference on Electrical, Electronics and Computer Science, Bhopal, 2014, pp. 1–6.
[7] K. Nikhil, M.K. Mishra, Application of hybrid energy storage system in a grid interactive microgrid environment, in: IECON 2015—41st Annual Conference of the IEEE Industrial Electronics Society, Yokohama, 2015, pp. 2980–2985.

[8] K. Nikhil, M.K. Mishra, Battery/supercapacitor based grid integrated microgrid with improved power quality features, in: Annual IEEE India Conference (INDICON) New Delhi, 2015, pp. 1–6.

[9] R. Sathishkumar, S.K. Kollimalla, M.K. Mishra, Dynamic energy management of micro grids using battery super capacitor combined storage, in: Annual IEEE India Conference (INDICON), Kochi, 2012, pp. 1078–1083.

[10] G. Wang, M. Ciobotaru, V.G. Agelidis, Power smoothing of large solar PV plant using hybrid energy storage, IEEE Trans. Sustain. Energy 5 (3) (2014) 834–842.

[11] Z. Zheng, X. Wang, Y. Li, A control method for grid-friendly photovoltaic systems with hybrid energy storage units, in: 4th International Conference on Electric Utility Deregulation and Restructuring and Power Technologies (DRPT), Weihai, Shandong, 2011, pp. 1437–1440.

[12] D. Yadeo, P. Chaturvedi, H.M. Suryawanshi, D. Atkar, S.K. Saketi, Transistor clamped dual active bridge converter to reduce voltage and current stress in low voltage distribution network, Int. Trans. Electr. Energ. Syst. 31 (2021), e12665.

[13] D. Yadeo, P. Chaturvedi, J.S. Lai, H.M. Suryawanshi, S.K. Saketi, A T-type dual active bridge with symmetrical configuration for solid state transformer, Int. J. Electron. 108 (12) (2021) 2019–2038.

[14] D. Yadeo, P. Chaturvedi, S.K. Saketi, A new five level dual active bridge DC-DC converter for solid state transformer, in: IEEE International Conference on Power Electronics, Drives and Energy Systems (PEDES), 2018, 2018, pp. 1–5.

[15] V. Ankita, A. Vijayakumari, A reduced converter count solid state transformer for grid connected Photovoltaic applications, in: International Conference on Emerging Technological Trends (ICETT), Kollam, 2016, 2016, pp. 1–7.

[16] L. Heinemann, G. Mauthe, The universal power electronics based distribution transformer, an unified approach, in: 2001 IEEE 32nd Annual Power Electronics Specialists Conference, Vancouver, BC, 2001, pp. 504–509.

[17] A.Q. Huang, Medium-voltage solid-state transformer: technology for a smarter and resilient grid, IEEE Ind. Electron. Mag. 10 (3) (2016) 29–42.

[18] C. Hunziker, N. Schulz, Solid-state transformer modeling for analyzing its application in distribution grids, in: International Exhibition and Conference for Power Electronics, Intelligent Motion, Renewable Energy and Energy Management, Nuremberg, Germany, 2016, pp. 1–8.

[19] Y. Liu, Y. Liu, B. Ge, H. Abu-Rub, Interactive grid interfacing system by matrix-converter-based solid state transformer with model predictive control, IEEE Trans. Ind. Inform. 16 (4) (2020) 2533–2541.

[20] W.A. Rodrigues, L.M.F. Morais, T.R. Oliveira, R.A.S. Santana, A.P.L. Cota, W.W.A. G. Silva, Analysis of solid state transformer based microgrid system, in: 12th IEEE International Conference on Industry Applications (INDUSCON) Curitiba, 2016, pp. 1–6.

[21] X. She, A.Q. Huang, R. Burgos, Review of solid-state transformer technologies and their application in power distribution systems, IEEE J. Emerg. Select. Top. Power Electron. 1 (3) (2013) 186–198.

[22] X. She, R. Burgos, G. Wang, F. Wang, A.Q. Huang, Review of solid state transformer in the distribution system: from components to field application, in: IEEE Energy Conversion Congress and Exposition (ECCE), Raleigh, NC, 2012, pp. 4077–4084.

[23] Sixifo, F., Xiaolin, M., Raja A., (2010). Topology comparison for solid state transformer implementation. IEEE PES General Meeting, Providence, RI, 1-8, 2010.

[24] Q.R. Muhammad, K. Abbas, A review on artificial intelligence based load demand forecasting techniques for smart grid and buildings, Renew. Sustain. Energy Rev. 50 (2015) 1352–1372.

[25] J. Blancas, J. Noel, Short-term load forecasting using fuzzy logic, in: 2018 IEEE PES Transmission & Distribution Conference and Exhibition—Latin America (T&D-LA), 2018, pp. 1–5.

[26] G. Priti, G. Monika, Short term load forecasting using fuzzy logic, in: International Journal of Engineering Development and Research (IJEDR) National Conference (RTEECE-2014), 2014.

[27] P. Ray, S.K. Panda, D.P. Mishra, Short-term load forecasting using genetic algorithm, in: H. Behera, J. Nayak, B. Naik, A. Abraham (Eds.), Computational Intelligence in Data Mining. Advances in Intelligent Systems and Computing, Springer, Singapore, 2019, p. 711.

[28] H. Mori, N. Kosemura, Optimal regression tree based rule discovery for short term load forecasting, in: Power Engineering Society Winter Meeting, IEEE 2001, 2001, pp. 421–426.

[29] B.J. Chen, M.W. Chang, C.J. Lin, Load forecasting using support vector machines: a study on EUNITE competition 2001, IEEE Trans. Power Syst. 19 (4) (2004) 1821–1830.

[30] M. Martín-Merino, J. Román, Electricity load forecasting using self organizing maps, in: S. Kollias, A. Stafylopatis, W. Duch, E. Oja (Eds.), Artificial Neural Networks—ICANN 2006. ICANN 2006. Lecture Notes in Computer Science, 4132, Springer, Berlin, Heidelberg, 2006.

[31] S.K. Dash, D. Patel, Short-term electric load forecasting using Extreme Learning Machine—a case study of Indian power market, in: IEEE Power, Communication and Information Technology Conference (PCITC), 2015, pp. 961–966.

[32] A.M. Tudose, I.I. Picioroaga, D.O. Sidea, C. Bulac, V.A. Boicea, Short-term load forecasting using convolutional neural networks in COVID-19 context: the Romanian case study, Energies 14 (2021) 40–46.

CHAPTER THIRTEEN

Machine learning and predictive control-based energy management system for smart buildings

Sachidananda Sen[a], Dharmendra Yadeo[a], Praveen Kumar[b], and Maneesh Kumar[c]

[a]Department of Electrical and Electronics Engineering, SR University, Warangal, Telangana, India
[b]Department of Electronics and Communication Engineering, VIT-AP University, Vijayawada, India
[c]Department of Electrical Engineering, Yeshwantrao Chavan College of Engineering, Nagpur, India

1. Introduction: Smart cities and smart buildings

Currently, in the year 2020, around 70% of population in the developing countries (like India and China) live in villages and towns. The remaining 30% people stay in the Tier I and Tier II cities. In this technological era, both online (internet) and offline (rail and road) connections are being established between the rural and urban population at an unprecedented speed. As cities offer better livelihood opportunities and lifestyle, a huge number of youth from the rural areas are looking toward cities as to make a living. This leads to rapid urbanization of cities as each day thousands of rural population migrate to the cities. This everyday addition of men power makes the cities highly suitable for economic activities like production of goods and services. With these benefits in such cities, there also arise concerns about overcrowding, ways for accommodating the ever-growing large population. Fulfilling the basic human needs like food, shelter, clothing, education, transportation, healthcare, water supply, electricity, sanitation, waste management, etc., puts huge burden on the existing unplanned conventional cities to operate satisfactorily [1].

The concepts of smart cities are introduced to sustainably accommodate the large and growing population in such cities. Various constituents or goals within the smart cities are identified, viz., smart buildings, smart industries, smart waste management, smart transportation or mobility systems, smart healthcare, etc., to gradually transform a conventional city to

Artificial Intelligence and Machine Learning in Smart City Planning
https://doi.org/10.1016/B978-0-323-99503-0.00015-6
199

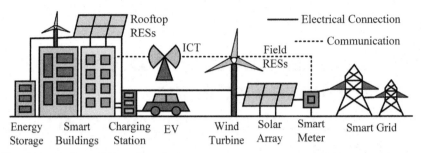

Fig. 1 Smart city and its various features.

a smart one [2]. Some of the smart cities features are given in Fig. 1. Among all these components, peoples spend most of their time inside a building whether at home or at workplace. Also, buildings consume most of the energy. Hence, it should be given special importance in realizing the features of a smart city. Now, a city comprises several buildings like residential apartments, commercial complexes, corporate offices, etc. Therefore, to increase the smartness of the city, its constituent buildings' smartness must be enhanced first. This can be achieved by including the residing humans, energy systems, monitoring units, and security systems as an integrated part of the building.

This book chapter provides a brief state-of-the-art literature on the smart buildings as important "building blocks" for realizing the vision of smart cities. It explains various components of the building instrumentation, energy systems, and automation, challenges, and future trends.

2. Energy management system for a smart building

A building can have various features or areas where it can be improved by incorporating different technologies. Some of these features are energy production and consumption, water recycling and rain water harvesting, biomass or solid waste management, ease of living and enhanced comfort level of the residents, monitoring and security, etc. [3]. From all these objectives of smart building, the energy management system (EMS) that regulates the energy production and consumption within, as well as power sharing with the external utility grid is the most critical one. The energy production is mostly done via renewables like solar photovoltaics (PV), active dynamic windows (building envelop), mini wind turbines, biomass, energy storage systems (ESS), backup diesel generator, etc. A smart building has to manage various loads like lighting, laundry rooms with washing machines, plug-in electric vehicles (EVs) charging station in smart parking lots, heating, ventilation, and air-conditioning (HVAC) or comfort management system for residing peoples, motors (elevator, escalators), pumps, etc. Furthermore, its integration to the smart grid is done to increase the reliability of power supply and for economic benefits by selling excess power when the generation is surplus [4].

This arrangement of various components working with different technologies makes us to consider a smart building as a system of systems (SoSs) architecture having green cyber-physical systems (CPS) that are basically interacting or interconnected networks of co-engineered physical and computational constituents. The main goal of the EMS is to minimize the power usage from the utility grid by having overall energy generation and storage equal to the total load of the building on any particular day. This aspect of smart building is also known as nearly zero energy building (nZEB), where a building is made self-sufficient in terms of its energy requirements [5]. Now, a building works as a small grid having its own sources and loads, which are known as microgrids (MGs). As the complete MG is attached to the whole building, it is named as building-integrated microgrids (BIMGs) [6]. Important components of smart building are shown in Fig. 2.

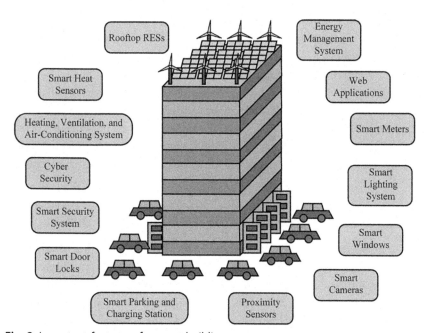

Fig. 2 Important features of a smart building.

2.1 Building-integrated microgrids

From an electrical engineering point of view, a smart building with its own generation and loads can be called as building-integrated microgrids (BIMGs). As a high-rise smart building has total load capacity much greater than its total generation capacity, connection of smart grid (SG) is inevitable. Therefore, from the EMS perspective, a smart building is basically a grid-connected microgrid (MG) where their interaction and power exchange are monitored using smart meters. An MG can operate in two modes, viz., grid-connected and islanding mode [7]. Hence, there has to have a proper protection and islanding detection arrangements to avoid any equipment damage. In addition to that, all the components within the BIMG require stringent control systems for optimal operation in an energy-efficient manner. For the task of monitoring, protection, and control of BIMGs, it is necessary to place various sensors and relays at specific locations for providing the right feedback to the controller units [8].

Furthermore, a smart building has several uncertainties to take care of like the weather variations, load forecasting, generation profiles of renewables, status of stored energy, grid power pricing, human occupancy and their varying behavior, different level of comfort requirements for each individual, etc. In control theory literature, predictive control strategies are applied to such systems with uncertainties to achieve the multiple objectives [8]. Various machine learning (ML) tools can be implemented to forecast these uncertainties accurately, which are subsequently fed into the predictive control algorithms to accommodate these uncertainties and obtain an optimally efficient EMS. Such control systems in literature on smart buildings are also referred as energy and comfort management system (ECMS) or building management system (BMS) [9]. It should have a built-in network of different sensors and actuators for implementing different control objectives and decision-making in automatic monitoring and control of building functions.

3. Predictive control-based EMS design for BIMGs

The need for an efficient EMS for smart building arises due to involvement of different types of sources, storages, and loads along with the respective uncertainties associated with them. To design the EMS, it is important to build the power balance model of the building. Now, modeling of the whole BIMG can be developed by considering the models of each components individually and aggregating them during formulation of the objective function. An objective or cost function is need to be defined to obtain the desired outcomes from the EMS. Some of the important objectives of an EMS for the BIMG are (1) getting maximum power output from the renewables like solar PV and mini wind turbines, (2) decreasing the number of charging and discharging cycles for the battery storage unit, (3) minimizing the power intake from the upstream utility grid, (4) providing uninterrupted supply to critical loads, (5) shifting of noncritical loads from peak hours to off peak hours, (6) reducing the operating cost of the energy system, etc.

Due to the presence of the system uncertainties, an EMS should be able to include the forecasting data of various power generation units and demand variation using the statistical information recorded prior to the day under consideration. The status or state of charge available in the storage devices and energy pricing readings from the smart meters during the peak and off-peak hours should also be incorporated. Along with the system uncertainties, various constraints of individual components are needed to be maintained within the allowable limits. In control literature, the most suitable controller to accommodate both uncertainties and constraints is the model predictive control (MPC). A brief discussion on the basics of MPC and its applications on the uncertain energy management systems within smart buildings is present here.

3.1 Fundamentals on model predictive control (MPC)

Recently, many complex power networks and power electronics modules are needed to be controlled by including some forecasting data as well as system-level constraints. With the presence of renewable energy sources (RESs) that has a limited amount of power generation, the need for power balancing and storage along with satisfying multiple system constraints within a BIMGs calls for a dynamic EMS. The MPC was developed in the 1980s and rigorously used in advanced process controls like chemical plants and oil refineries. The main advantage of using MPC is that it accounts for the future timeslots for optimizing the present timeslot [10]. It has the capability to forecast the future events and include them into the current control action. It is an optimal control scheme that works in an iterative manner, i.e., it optimizes the finite time-horizon over a predefined predictive horizon, but it incorporates only the current timeslot over the control horizon and then repeats the optimization process.

Therefore, theory of the MPC method is mainly based on iterative, finite-horizon optimization of a plant/system model. At time instant t, the present states of the plant are sampled and a cost or objective minimizing control scheme is calculated using a numerical minimization algorithm for a relatively short future time horizon: $[t, t+T]$. In most of the cases, online computation is used to obtain the state trajectories that originate from the current state and determine (using the solution of Euler–Lagrange equations) a cost-minimizing control scheme till the predictive time-horizon $t+T$. Thereafter, only the first step of the obtained control law is implemented, and then, the plant/system state is sampled once again and all the calculations are repeated starting from the latest current state, yielding a new control action and new predicted path of the state dynamics. Due to the forward shifting of the predicted horizon, the MPC technique is also known as the receding horizon control that directs the system dynamics to follow a desired trajectory where the objective/cost function is minimized. Even though this approach is not always optimal, but in practice, it has shown very good results on many industrial processes [11].

The MPC is mostly a multivariable controller design algorithm that mainly considers three constituents: (1) an internal dynamic model of the system or process, (2) a cost/objective function J defined as deviations in controlled and reference variables over the receding horizon, (3) an optimization algorithm to minimize the objective function J using the control input u. A typical example of a cost or objective function for optimization is represented as (1)

$$ J = \sum_{i=1}^{N} w_{x_i} (r_i - x_i)^2 + \sum_{i=1}^{N} w_{u_i} \Delta u_i^2 \tag{1} $$

subject to certain minimum and maximum limits of system constraints. Here, r_i, x_i, and u_i are the reference, controlled, and manipulated variables, respectively. Also, w_{x_i} and w_{u_i} are known as the weighting coefficients to reflect the relative importance to states x_i and penalty factor for a large change in control input u_i, respectively. It is important to understand that a predictive control involving some form of forecasting and optimization of an objective function with defined constraints also broadly falls under the general MPC algorithm.

3.2 Developing models of BIMGs

A building can have some critical loads requiring uninterrupted power supply along with the noncritical ones, for example, security systems like CCTV cameras, lighting of entry/exit points, stairs, and corridors, control room and datacenter, dispensary or healthcare unit, etc. The main objective of an EMS is to provide power to these critical loads at utmost priority. Therefore, when the BIMG is operating in islanded or isolated mode (due to supply interruption from main grid), depending upon the generation from renewables and the state-of-charge (SoC) of the batteries, the non-critical loads are removed and only critical loads are provided with electricity.

An application for modeling smart building microgrid to develop an objective function, and its subsequent optimization for realizing BEMS that uses machine learning (ML)-based prediction data has been demonstrated in later section.

4. Smart homes

A smart building can have multiple utility units like, common hall area, office spaces, working stations, laundry rooms, gym, individual homes/flats, etc. Each home can have different number of people with different level of comfort requirements, as well as electrical and electronic goods like refrigerator, fans, geysers, microwave oven, television, etc. Therefore, each smart home must be handled as separate units or systems inside a smart building. This leads to a separate research area where all these equipment can be controlled using computer or mobile phones in an interconnected manner known as internet of things (IoT).

At an individual level, each apartment can also be efficiently managed as smart homes. Here, various digital appliances, viz., lighting, refrigerators, air-conditioners, heating, windows or ventilation control, door locking, and security systems, are connected and communicated seamlessly [12]. All these components can be controlled through mobile applications (apps) or computer and provide the occupant/user an access or interface to remotely operate these electronics.

A discussion on smart homes technology (SHT) having interconnected appliances with internet of things (IoT), information and communication technologies (ICT), and decision-making as well as control algorithm design using the field programmable gate array (FPGA) is also included. FPGA has several advantages when it comes to setting up IoT-based home appliance system where coordinated decision-making is crucial [13]. Different electronics have different manufacturers, therefore, interfacing difficulty arises. Various networking and communication protocols need to be followed.

4.1 Basics of FPGA

The FPGA is an electronic device that has important applications in the area of power system, smart grids, and communication technologies that need the features of re-configurability, high bandwidth data transfer, parallel processing, highly efficient computational capabilities, and low latencies [14]. By the implementation of FPGA, signal processing, data conversion from analog-to-digital (ADC) and digital-to-analog (DAC) modules, and lookup table-based preset decision making are possible with a higher degree of accuracy that too in a cost-effective manner. In addition to that, FPGA chips are easy to use, having reprogrammable benefits that assist in accommodating any adaptation or change in the design process [15]. The efficiency and performance of the FPGA-built systems are by far better than the other computational equipment like the microcontroller, i.e., μC and digital signal processing (DSP)-based sensors and relays. As the FPGA-developed devices work on the principle of parallel processing, its operational abilities are highly efficient and faster [16]. The aforementioned predominant features make the FPGA-designed systems highly suitable for performing different applications like power flow monitoring, identification of various faults, protection of the smart grid, controlling different components of smart building, and developing IoT-based smart home appliance system [13].

4.2 Role of FPGA is developing IoT for smart homes

The smart homes are the concept of 21st century where the internet of things (IoTs)-enabled homes with all the household appliances, i.e., automated-coffee maker to smart water management, are automated and controlled by IoT-enabled technology. To make the home appliance smart or make a building automated, we need to add features like artificial intelligence, communication technology, data processing, big data analysis, etc., to the appliance [17]. Therefore, in the home automation model, all the household appliances must be connected through a centralized controller through a wireless technology with bidirectional communication that can take place between the central controller and different types of physical devices through the IoT sensors. The sensor can sense the real-time condition of the end appliances in the form of data that can be sent to the central controller, where decision can be received and after authentication access of the particular appliances can be provided to the user. Therefore, a field programmable device is required, having features like re-programmability, re-configurability, low latency, high speed, fast response, easy to interface with the communication protocol and also able to interface IoTs sensor, etc., to create a fully automated smart system [18]. A basic layout of a smart home is shown in Fig. 3. In this layout, different types of sensors are part of a smart home, through which physical devices can be connected. A real-time face recognition or any other mode for authentication can be interfaced to provide better security. Real-time data can be transferred to the central controller through the wireless channel/module. At the central controller, data can be processed and based on the condition access can be provided. Here, a cloud is proposed to use for the data analysis and these data can also be used for the development of artificial intelligence-enabled environment.

There are different areas of research possible that can be explored and implemented using the FPGA as a central processing unit.

Interfacing of different types of sensor to actuate different types of controlled actuators:

(a) Enabling of smart meter to provide better efficient home energy management system.

(b) Self-controlled window system of smart homes.

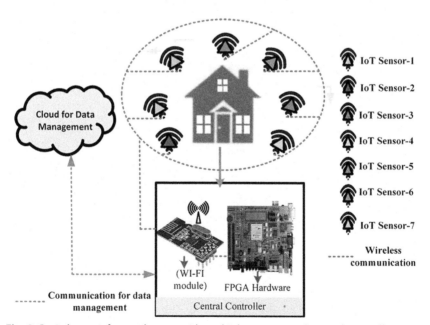

Fig. 3 Basic layout of smart homes with multiple sensors and central controller.

4.3 IoTs for home energy management (HEM) using FPGA

One of the key challenges in a smart building is that how to optimize energy usage. In a smart home, each and every device is being operational by the use of electric power. Also, smart meter is an important device that can be used to receive and transmit the power to and from the utility grid. Since, most of the buildings are equipped with the green and RESs, therefore, each and every device can be controlled by the central controller for optimal use of the energy [19,20]. The cost of the energy depends on the demand and supply of over the generated energy. The amount of consumed energy during the peak hours can be reduced or optimized, by turning off the nonessential devices that can be used when the demand of the electricity is low. Therefore, the end user devices can be divided into several groups based of their uses. The grouped devices (essential and nonessential) can be controlled by using the FPGA-based central controller.

Another aspect of the smart energy management is to transfer the excess energy generated by the building RESs during the peak hour to the grid when the price of electricity is more. FPGA can be used as a central controller to support and develop an efficient energy management environment from which each and every device is connected via a wireless network. Fig. 4 shows an idea of smart energy management system through which refrigerator, water heater, microwave oven, and charging point of the electric vehicle. Each of the nodes or end devices can be connected through the

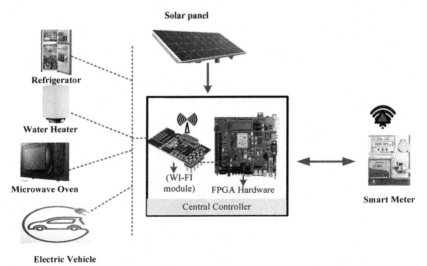

Fig. 4 Concept of IoT based energy management system for smart homes.

FPGA-based central controller. These devices can be divided into two groups and can be operated into different tariffs. Also, solar panel is connected and controlled by the FPGA to manage the power flow from or to the utility grid. As per the requirement, end devices can be added and removed from one to another group. This can be done without changing the hardware, since FPGA supports re-programmability; by using this feature, we can add more number of devices without changing much of the hardware [16].

5. Application of machine learning

In this section, an application of machine learning (ML) in forecasting various time-varying parameters from both load and generation sides, which is to be used in predictive control, is presented. ML is an efficient tool to forecast or estimate the future values of different variables (at present time) by using the past statistical database. There are various algorithms and set of rules to be followed to apply the ML technique.

5.1 Brief description on artificial intelligence (AI) and machine learning (ML)

Artificial intelligence (AI) can be defined as the intelligence demonstrated by the machines as a software program, which might resemble the natural intelligence displayed by humans or animals. It is a field of study of "intelligent agents or codes" that are defined as the system that perceives its environment and takes actions to maximize its chance of attaining desired goals [21].

Some of the best AI applications and examples are advanced web search engines (i.e., Google), recommendation or suggestion systems (found in Amazon, YouTube, Facebook, and Netflix), understanding and recognizing human voice and acting on a task (like Siri or Alexa), self-driving or driverless cars (as developed by Tesla), and competing at the highest level in strategic gaming systems (e.g., chess and Go). The field of AI is quite fast and it includes, machine learning, data mining, fuzzy logic systems, evolutionary optimization, and multiagent systems (MAS). It is important to understand that ML is a part or subset of AI.

Now, coming to the ML, it is the study of computer algorithms that can improve automatically or are self-improving, through experience, training, and gaining knowledge by the use of data. Initially, ML algorithms build a model based on the available sample data, which is known as "training data," to be able to make predictions, differentiate, or take decision without being explicitly programmed for it. It is used in a wide variety of applications, such as in medicine, forecasting, email filtering, prediction of stock prices, speech recognition, and computer vision, where it is unfeasible or somewhat difficult to develop using the conventional algorithms to complete the required tasks.

In this regard, computational statistics is also a subset of ML that involves in making predictions using computers. However, not all the ML can be considered as statistical learning. Data mining is a field that focuses on exploratory data analysis by applying unsupervised learning techniques. Other applications of ML use data sets and neural networks in a manner to mimic the working of a biological brain. Because of its implementations across multiple areas like history, future, medical science, business problems, ML is also known as predictive analytics.

5.2 Application of ML on the EMS of a smart building

The smart building should have an energy efficient structure with a zero energy waste concept. These buildings can also conceptualize with the term "prosumers," as it can simultaneously produce and consume the energy. An optimal energy management system (EMS) for these buildings can be realized in various manners. With respect to the available loads and existing sources, the building energy management system (BEMS) can be implemented. In this chapter, renewable energy sources (RESs) and battery storage-based energy management for the smart buildings is proposed. It is somewhat similar to the EMS of microgrids [22,23].

Methodology:

A smart building with **D** kW of connected load is considered. This building has a battery storage room and small-scale rooftop-based RESs such as solar photovoltaic (SPV) and small-wind turbine units. The SPV modules can also be used in the form of smart window structures of the building that adjusts itself to the optimal direction of solar irradiance. The building is grid-connected, i.e., it also has the access to get supply form the main grid. Since the renewable energy and the load are considered to be stochastic in nature, an appropriate forecasting methodology should be implemented [24]. The RES and the load demand forecasting have been incorporated using machine learning (ML) techniques and a day-ahead optimal dispatch of various energy producing sources has been obtained through a nonlinear optimization algorithm. Subsequently, based upon the results of optimal dispatch from various sources, the BEMS is implemented (Fig. 5).

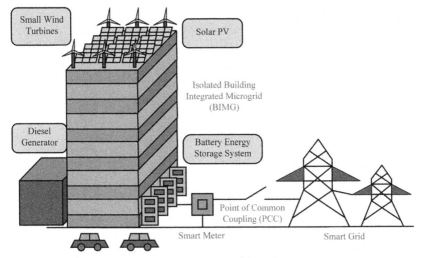

Fig. 5 An islanded Building Integrated Microgrid (BIMG).

The specifications of the overall system are as follows:
The building total load demand **D** = 100kW,

- **SPV system specifications**
  1. RC_{SPV} (kW) = 0.25 kWp (rated capacity of solar panel)
  2. O_{PV} = SPV system output (kW)
  3. Derating factor DF_{SPV} = 10%
  4. H_{std} = 1 kW/m² (standard solar irradiance on a surface)
  5. H = solar irradiance on a particular surface
  6. Temperature coefficient "θ" (°C) = 0.5
  7. T_{stc} = 20°C (Standard air temperature)
  8. T_c = cell temperature (°C)
  9. T_a = average ambient cell temperature (°C)
  10. NOCT = nominal operating cell temperature (43°C)

$$O_{PV} = RC_{SPV}{}^* DF_{SPV}{}^* \frac{H}{H_{std}} \left(1 + \theta^*(T_c - T_{stc})\right) \qquad (2a)$$

$$T_c = T_a + \frac{H}{0.8}(NOCT - 20) \qquad (2b)$$

- **Wind turbine unit specification**
  1. C_p (%): 40 (coefficient of maximum power)
  2. O_{wg}: wind output (kW)
  3. ϱ_a (lb = ft3): 0.06841 (air density)

4. A_r (ft2): 684.423 (rotor swept area)
5. k_{wg}: 0.000133 (constant to obtain the output power)
6. v (mph): (wind speed)

$$O_{wg} = \frac{1}{2} * k_{wg} * C_p * \rho_a * A_r * v^3 \tag{3}$$

- **Battery storage system (BSS) specifications**
1. C_{max}: 3 kWh (maximum charging capacity)
2. η_{ch}: 0.95 (charging efficiency)
3. η_d: 0.95 (discharging efficiency)
4. RC_{bss}: 2 kW (maximum output)
- **Diesel generator (DG) specifications**
1. RC_{dg}: 5 kW (rated capacity)
2. ϱ_{dg}: 0.264 (DG fuel cost $/kWh)
3. C_{tax}: 0.04 (CO_2 emission tax ($/kg))
4. C_{in}: 0.758 (CO_2 intensity kg/kW)

5.2.1 Objective function

It is an operational cost-based function that includes various operation costs pertaining to building integrated microgrid (BIMG) and need to be minimized. These costs include fuel cost associated with diesel generator (DG), cost associated with desired loss of load in the system, cost in terms of penalty for the CO_2 emission, and cost associated with losses in the storage unit.

$$\min (f) = \frac{1}{s} \left(\sum_t \sum_s C_{dl} * D_{dl}(t, s) + FC_{dg} \left(\sum_t \sum_s O_{dg}(t, s) \right) \right. \tag{4}$$

$$\left. + \left(\sum_t \sum_s O_{dg}(t, s) * C_{tax} * C_{int} \right) + \sum_t \sum_s (C_{d,op}) \right)$$

where

C_{dl}: cost associated with desired loss of load or unserved load ($)
D_{dl}: unserved load demand (kW)
C_{dg}: fuel cost associated with the diesel generator
O_{dg}: output of the diesel generator
C_{tax}: penalty or tax associated with the CO_2 emission
C_{int}: carbon intensity

$C_{d,op}$: daily operating costs associated with various DERs

t: time index

s: scenario index

The daily operating costs include the cost associated with turning on/off of DG unit, cost associated with operation and maintenance of DERs, and cost associated with loss in storage system, etc.

5.2.2 Important system constraints

Some of the equality and inequality system constraints to be satisfied are:

(a) Load–generation constraint

$$D(s, t) = N_{PV}*O_{PV}(s, t) + N_{wg}*O_{wg}(s, t) + O_{dg}(s, t) + O_{bat}(s, t) + D_{dl}(s, t) - C_p(s, t) \tag{5}$$

where $D(s,t)$ is the total load demand, N_{PV} and N_{wg} are the number of solar PV and wind turbine units, $C_p(s,t)$ is the capacity of the storage system.

(b) Battery energy storage system (BESS) constraint

$$C_{eng}(s, t) \leq N_{bess}*C_{max} \tag{6}$$

where $C_{eng}(s,t)$ is the BESS stored energy and N_{bess} is the number of storage units.

(c) Diesel generator (DG) output constraint

$$O_{dg}(s, t) + SR(s, t) \leq N_{dg}*RC_{dg} \tag{7}$$

where $SR(s,t)$ is the spinning reserve and N_{dg} is the number of DG units.

(d) Simultaneous charging/discharging constraint of BSS

$$C_p(s, t)*O_{bat}(s, t) = 0 \tag{8}$$

(e) Renewable energy penetration constraint

$$\sum_{s}^{S} \sum_{t}^{T} N_{wg}*O_{wg}(s, t) + \sum_{s}^{S} \sum_{t}^{T} N_{PV}*O_{PV}(s, t)$$
$$\geq PP* \sum_{s}^{S} \sum_{t}^{T} (D(s, t) - D_{dl}(s, t)) \tag{9}$$

where PP is the percentage penetration of RESs.

(f) Reliability constraint

$$US_{max} \geq \frac{\displaystyle\sum_{s}^{S} \sum_{t}^{T} D_{dl}(s, t)}{\displaystyle\sum_{s}^{S} \sum_{t}^{T} D(s, t)} \tag{10}$$

where US_{max} is the ratio of the unserved load demand to the total load demand at any time "t" and scenario "s"

5.2.3 Renewable energy and load data prediction

The actual and the predicted data corresponding to a given set of wind speed and time of the day as an input to obtain the wind power output is shown in Fig. 6. The decision-tree learning approach is implemented to find the 24-hr prediction model of the wind output power in MATLAB environment. This method is considered to be a powerful tool in data mining, statistics, and machine learning. In data mining, decision trees can also be described as the combination of mathematical and computational techniques to aid the description, categorization, and generalization of a given set of data [25].

The obtained training model has following characteristic:

Root mean square error (RMSE): 6.028e−05, R-squared value: 0.98, mean square error (MSE): 3.634e−09, mean absolute error (MAE), 2.296e−05, training time: 78.69 s.

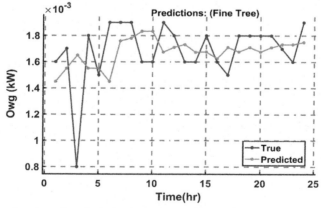

Fig. 6 Actual and predicted wind output power data under consideration.

The actual and the predicted solar power output data for a given set of solar irradiance over a given geographical area of Roorkee, Uttarakhand, India, are shown in Fig. 7. The predicted data for SPV output are obtained through the rational quadratic Gaussian process regression (GPR) method, which is used to train regression models to predict data using supervised machine learning. Here we take a training data set of the form (x_i, y_i); $i = 1, 2, 3, ..., n$, where $x_i \varepsilon R^d$ and $y_i \varepsilon R$ [26]. The forecasting is done by considering the time of the day, humidity, and temperature as input variables. The 10% of them are used for testing purposes.

The obtained forecasted model parameters are as follows:

RMSE: 21.92, R-squared value: 0.99, MSE: 480.73, MAE: 11.492, and training time: 66.419 s.

The actual and the predicted data of load demand for a typical 24-hr period have been shown in Fig. 8. Beta distribution has been utilized to model the load connected to the smart building MG system, which is a continuous

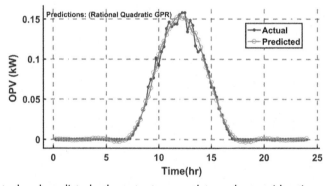

Fig. 7 Actual and predicted solar output power data under consideration.

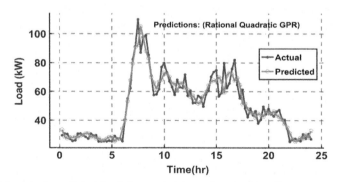

Fig. 8 Actual and predicted load demand data under consideration.

probability distribution, and the Gaussian process regression (GPR) method
[26] is used for the prediction as well.

The obtained model has the following characteristics:

RMSE: 6.75, R-squared value: 0.88, MSE: 45.57, MAE: 4.94, and
training time: 47.58 s.

For the present study and the analysis, the overall renewable generation and
the load demand data were considered for Roorkee area [27,28].

5.2.4 Methodology used for optimization

The sequential quadratic programming (SQP) approach is used to solve
many real-world nonlinear problems. SQP approach divides the nonlinear
problems into linear subproblems as it linearizes the available nonlinear
constraints. This approach is a powerful tool to handle any degree of nonlinearity in an optimization problem and in constraints. In general, an SQP
problem is formulated as below:

$$\min \ f(x)$$
$$\text{s.t}$$
$$h(x) = 0;$$
$$g(x) \leq 0$$

$$(11)$$

where $f(x)$, $h(x)$, and $g(x)$ are various system constraints. x is a vector.

5.2.5 Simulation results

Fig. 9 shows the optimal dispatch of DERs for a typical 24-h period with
respect to various reliability conditions (US_{max}) starting from 0% to 30%
for the SQP algorithm with "fmincon" as a solver. The reliability condition
shows a maximum percentage of desirable loss of load into the building MG
system.

The subfigures, i.e., (A), (B), (C), and (D) of Fig. 10, show the optimal
switching sequence of diesel generators over a typical 24-h period under various reliability conditions, viz., $US_{max} = 0\%$, 10%, 20%, and 30%, respectively, for the SQP approach. It can be seen that during the early hours
of the day, the energy requirement of the load is high. Also, during that
interval, the power availability from renewable sources such as solar is less.
Therefore, the optimal number of DG required is high during these periods
as compared to the period of high renewable power availability and is almost
nil during the 13th and 14th hour, i.e., during afternoon period.

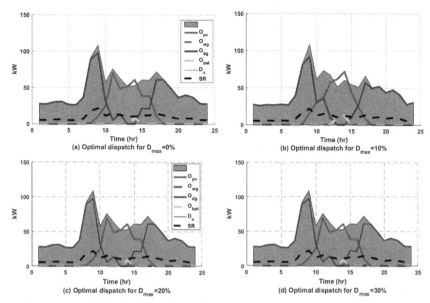

Fig. 9 A typical 24hr optimal dispatch of DRs with various US_{max} for the SQP algorithm. (A) For $US_{max}=0\%$, (B) for $US_{max}=10\%$, (C) for $US_{max}=20\%$, (D) for $US_{max}=30\%$.

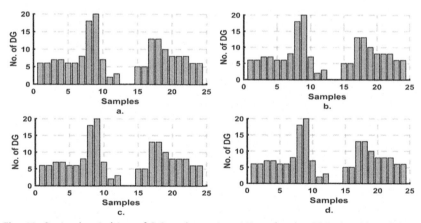

Fig. 10 Optimal switching of DG under various US_{max} for the SQP algorithm. (A) For $US_{max}=0\%$, (B) for $US_{max}=10\%$, (C) for $US_{max}=20\%$, (D) for $US_{max}=30\%$.

6. Future trends and research challenges in smart building

From the aforementioned discussions on smart buildings, there are several future trends and research topics that are required to be realized. Some of the important ones are presented here.

- Accurate sensing of the number of occupants and designing building control around the occupant centric needs.
- To maintain an uninterrupted supply for any critical loads (hospital, fire station etc.), the relays are required for fast switching from the grid-connected mode to the islanded mode. Here, the use of FPGA-based digital relays can be explored for islanding detection and achieving mode adaptability.
- Several data communications between different electronic equipment are needed in smart building. Here, instead of using the expensive wired communication infrastructure, more efforts toward the use of economical wireless communication modules could be given.
- Smart homes applications based on IoT should try to optimize the comfort and health, follow safety norms for electrical components using fire systems, provide security from external intrusion, and give privacy to the user as well as less predict ones behavior.
- Use of artificial intelligence and machine learning for both renewable generation and load demand forecasting techniques for BIMGs is needed to be further researched.

Considering the aforesaid future trends and research gaps for smart building, the technical challenges to be worked upon are summarized below:

- Data privacy and cybersecurity issues due to cyber-physical system have to be handled in the smart building infrastructure.
- Developing prototypes FPGA-based smart digital relays with fast sensing abilities, wireless communication with nearby sensors or relays, decision making capabilities, and having multiple functionalities [29].
- Achieving not only the target of nearly zero energy buildings (NZEB) but to become a positive energy building (PEB), i.e., energy surplus building. It calls for aggressive installation of RESs, water recycling, generating energy from biomass [30,31].
- Application of different domains of artificial intelligence, viz., machine learning, fuzzy logic, agent-based systems, evolutionary optimization, etc., for the energy management of self-sustainable existence of the smart buildings.

References

[1] R.K.R. Kummitha, "Smart cities and entrepreneurship: an agenda for future research," Technol. Forecast. Social Change, vol. 149, pp. 119763, October 2019.

[2] J. Shah, J. Kothari, and N. Doshi, "A survey of smart city infrastructure via case study on New York," Proc. Computer Sci., vol. 160, pp. 702-705, Nov. 2019.

[3] Al Dakheel J., Del Pero C., Aste N., Leonforte F., "Smart buildings features and key performance indicators: a review," Sustain. Cities Soc., vol. 61, pp. 102328, Oct. 2020.

[4] D.S. Shafiullah, T.H. Vo, P.H. Nguyen, A.J.M. Pemen, Different smart grid frameworks in context of smart neighborhood: a review, in: 52nd International Universities Power Engineering Conference (UPEC), Aug. 2017, pp. 1–6.

[5] M. Schmidta, and C. Åhlund, "Smart buildings as cyber-physical systems: data-driven predictive control strategies for energy efficiency," Renew. Sustain. Energy Rev., vol. 90, pp. 742–756, April 2018.

[6] H. Fontenot, B. Dong, Modeling and control of building-integrated microgrids for optimal energy management—a review, Appl. Energy 254 (2019) 113689.

[7] S. Sen, V. Kumar, Assessment of various MOR techniques on an inverter-based microgrid model, in: 14th IEEE India Council International Conference, 2017, pp. 1–6.

[8] S. Sen, V. Kumar, Microgrid control: a comprehensive survey, Ann. Rev. Contr. 45 (2018) 118–151.

[9] J. Bakakeu, F. Schäfer, J. Bauer, M. Michl, J. Franke, Building cyber-physical systems—a smart building use case, in: Smart Cities: Foundations, Principles and Applications, John Wiley & Sons, 2017, pp. 605–639. ch 21.

[10] M. Arnold, Model predictive control of energy storage including uncertain forecasts, in: 17th Power Systems Computation Conference, August 2011, pp. 1–7.

[11] P.O.M. Scokaert, D.Q. Mayne, Min-max feedback model predictive control for constrained linear systems, IEEE Trans. Autom. Contr., vol. 43, no. 8, pp. 1136–1142, August 1998.

[12] M. Casini, Active dynamic windows for buildings: a review, Renew. Energy (2017).

[13] S. Sharma, R. Deokar, FPGA based cost effective smart home systems, in: International Conference on Advances in Communication and Computing Technology, 2018, pp. 397–402.

[14] E. Monmasson, L. Idkhajine, M.W. Naouar, "FPGA-based controllers," IEEE Ind. Electron. Mag., vol. 5, no. 1, pp. 14–26, Mar. 2011.

[15] R. Dubey, P. Agarwal, and M.K. Vasantha, "Programmable logic devices for motion control—a review," IEEE Trans. Ind. Electron., vol. 54, no. 1, pp. 559–566, Feb. 2007.

[16] P. Kumar, V. Kumar, R. Pratap, Design and implementation of phase detector on FPGA, in: 6th IEEE International Conference on Computer Applications in Electrical Engineering-Recent Advances (CERA), 2017, pp. 108–110.

[17] K. Amleset, H. Gaber, IoT for home energy management (HEM) using FPGA, in: IEEE 9th Int. Conference on Smart Energy Grid Engineering (SEGE), 2021, pp. 54–57.

[18] R. Krishnamoorthy, K. Krishnan, C. Bharatiraja, Deployment of IoT for smart home application and embedded real-time control system, Mater. Today: Proc. 45 (2021) 2777–2783.

[19] P. Kumar, V. Kumar, R. Pratap, RT-HIL verification of FPGA-based communication-assisted adaptive relay for microgrid protection, Electr. Eng. (2021) 1–11.

[20] P. Kumar, V. Kumar, R. Pratap, Prototyping and hardware-in-loop verification of OCR, IET Generat. Transm. Distrib. 12 (12) (2018) 2837–2845.

[21] S. Wendzel, J. Tonejc, J. Kaur, A. Kobekova, Cyber security of smart buildings, in: Security and Privacy in Cyber-Physical Systems: Foundations, Principles, and Applications, John Wiley & Sons, 2018, pp. 327–357. Ch 16.

[22] S. Sen and V. Kumar, "Decentralized output-feedback based robust LQR V-f controller for PV-Battery microgrid including generation uncertainties," IEEE Syst. J., vol. 14, no. 3, pp. 4418-4429, Sept. 2020.

[23] S. Sen and V. Kumar, "Simplified modeling and HIL validation of solar PVs and storage based islanded microgrid with generation uncertainties," IEEE Syst. J., vol. 14, no. 2, pp. 2653-2664, June 2020.

[24] S. Sen, V. Kumar, Microgrid modelling: a comprehensive survey, Ann. Rev. Contr. 46 (2018) 216–250.

[25] R.S. Milln-Castillo, E. Morgado, and R. Goya-Esteban, "On the use of decision tree regression for predicting vibration frequency response of handheld probes," IEEE Sensors J., vol. 20, no. 8, pp. 4120–4130, April 2020.

[26] N. Zhang, J. Xiong, et al., Gaussian process regression method for classification for high-dimensional data with limited samples, in: IEEE International Conference on Information Science and Technology, 2018, pp. 358–363.

[27] M. Kumar, B. Tyagi, An optimal multivariable constrained nonlinear (MVCNL) stochastic microgrid planning and operation problem with renewable penetration, IEEE Syst. J. 14 (3) (2020) 4143–4154.

[28] M. Kumar, B. Tyagi, Multi-variable constrained nonlinear optimal planning and operation problem for isolated microgrids with stochasticity in wind, solar, and load demand data, IET Gen. Trans. Dist. 14 (11) (2020) 2181–2190.

[29] P. Kumar, V. Kumar, R. Pratap, Digital design and implementation of an overcurrent relay on FPGA, in: 14th IEEE India Council International Conference, 2017, pp. 1–5.

[30] S. Sen and M. Kumar, "MPC based energy management system for grid-connected smart buildings with EVs," IEEE IAS Global Conference on Emerging Technologies, May 2022. (Accepted).

[31] M. Kumar, S. Sen, and S. Kumar, A robust performance analysis of a solar PV-battery based islanded microgrid inverter output voltage control using dual-loop PID controller, IEEE IAS Global Conference on Emerging Technologies, May 2022. (Accepted).

CHAPTER FOURTEEN

Effective prediction of solar energy using a machine learning technique

B. Vikram Anand, G.R.K.D. Satya Prasad, and Bishwajit Dey
Department of Electrical and Electronics Engineering, GIET University, Gunupur, Odisha, India

1. Introduction

The 21st century is defined by rising energy use and greenhouse gas emissions, which are leading to changing climate in unprecedented ways. Fossil fuels continue to be our primary source of energy and heat generation, responsible for 42% of greenhouse emissions in 2016 [1]. Energy efficiency and the growth of renewable energy sources are offered as the two primary ways to reduce these emissions [2], with photovoltaic solar energy becoming one of the speediest renewable energy sources due to its cheap maintenance and operating costs [3]. Solar panels may also generate electricity everywhere there is adequate sunshine without having a negative influence on the environment, allowing for energy production in congested situations such as townships or industrial parks.

To calculate the solar energy potential of a building's rooftop, we need two things: (1) number of solar modules may be installed on such a roofing section as well as (2) how much energy each module can produce in a year, taking into account local irradiance and shading. As shown in Fig. 1, the current study presents a framework for answering these issues.

The solar energy assessment procedure may be expensive and time-consuming, requiring anything from an hour to two full days to determine each rooftop's solar potential.

This has culminated in the expenses and sales accounting for up to 30%–40% of overall project expenses in the solar sector, greatly hurting the unit arithmetic of solar installations.

Through using deep learning to automate these analyses, our research hopes to substantially lower the cost of this procedure and provide this information easily accessible to both building owners and solar energy businesses.

Artificial Intelligence and Machine Learning in Smart City Planning Copyright © 2023 Elsevier Inc.
https://doi.org/10.1016/B978-0-323-99503-0.00019-3

Fig. 1 Detailed workflow for the proposed idea.

The primary goal of this effort is to forecast potential solar resources of residential and nonresidential rooftops.

In reality, it would be essential to massively calculate the roof's topology and weather conditions, as well as certain other critical components such as the surface area exposed for solar module installation; the tendency and orientation, which seem to be important components in the production of energy via solar photovoltaic panels; and thorough sunlight dimensions, guess it depends on the geolocation. However, this knowledge may be provided at the municipal level, but it is seldom available at the national level.

We employed a machine learning algorithm trained on multiple datasets to forecast missing information and enable large-scale estimates. We use open data to (1) train machine learning models, (2) detect rooftops in aerial pictures, and (3) forecast their inclination, allowing us to (4) estimate the total amount of mountable modules plus their (5) electricity production.

2. Significance of this estimate

We need to take a step back and examine the background before diving into the answer. Meeting energy consumption while minimizing global warming will be a huge issue in our futuristic society of 10 billion people. Nonetheless, fossil fuels continue to be our primary source of power and heat production, accounting for more than 40% of global greenhouse gas emissions in 2016 [1]. The two primary options for lowering these figures are energy efficiency and the growth of renewable energy.

Despite the inclusion of nonrecyclable elements and relatively low performance of 15% compared to 50% to 90% for hydroelectric or wind power [2], photovoltaic arrays are among the fastest growing renewable energy. However, their operating and maintenance expenses are cheap, and modules may be put on the ground as well as on top of a roof, which is advantageous in cities where space is limited. Estimating large-scale solar panel output is thus a critical problem for accelerating this shift.

3. Research technique

As shown in Fig. 2, the current study presents an approach for answering these issues. First, an imagery segmentation method applied to aerial footage is utilized to extract the 2D geometry of each roof section and any obstructing objects it may include. A mix of geometric approaches as well as a random forest algorithm generates the roof's 3D geometry

Fig. 2 Five steps to predict solar potential.

(pitch/azimuth). Its maximum number of parameters that can fit on a section is then calculated using a geometrical packing procedure. A shading mask is calculated based on the shadows produced by trees or buildings and relief, and the particular photovoltaic power is calculated based on climatic data, module orientation, and shading effects.

3.1 Object and roof segmentation

The initial phase in our process is to partition roof pieces and the equipment atop them. This will assist us in determining how much accommodation is on a rooftop for solar panels. We divided this job into understanding segmentation tasks, each of which consisted of categorizing every pixel of an object into one of many classes [4]. As shown in Fig. 3, we constructed one model that segments the photographs into background, roof sections, and roof ridges, and another that involves dividing the images into background and a collection of selected roof objects. Geometrical regularization improves the eventuality of roof portions.

3.2 Calculating the azimuth and pitch of a roof

Having the correct roof surface limits allows us to calculate the azimuth geometrically (orientation). However, extracting detailed information from aerial pictures, such as the pitch (inclination) of a slope, is difficult.

We've decided to geometrically generate inclination parameters from the 3D training sample and use them as identifiers to train a random forest regression model [4] to produce the pitch [5]. The elements we utilized to characterize the roof and feed the model that incorporates building-centered valuable information, including the footprint area, circumference, number of surrounding buildings, roof material and kinds, and floor number.

3.3 Increase the quantity of solar panels

We knew the roof slope borders, their surface, direction, and inclination, as well as the possible blocking items up to this point as seen in Fig. 4. To determine the greatest number of solar panels that might fit into the roof, we simply packed as many solar panels as possible along the major axis of the design [6]. Then, we deleted panels that intersected with items or formed roof limits.

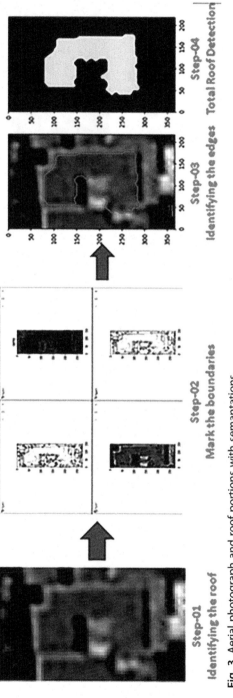

Fig. 3 Aerial photograph and roof portions with semantations.

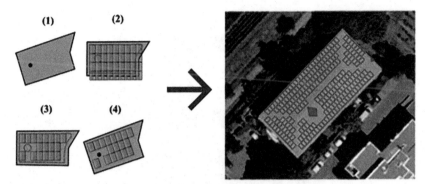

Fig. 4 The packing algorithm is illustrated as follows: (1) determine the main axis of the roof slope; (2) eagerly insert panels; (3) delete panels overlapping objects or borders; and (4) calculate the enclosed area. The illustration shows two roof items as an example.

3.4 Calculating solar potential

A location's solar potential is defined as the quantity of solar energy it can receive in a year. Solar energy (or solar power) is a type of power such as heat or electricity that may be converted using various methods, such as solar panels.

However, not all solar panels produce the same amount of solar electricity. It is affected by field, azimuth, whole irradiation (global level and direct normal irradiation), and geolocation. It is defined as PV output (kWh/kWp): the installation's particular photovoltaic power output, an average of the yearly production normalized to 1 kWp of installed capacity. It may be computed using sites such as PVGIS and the Global Solar Atlas.

For our purposes, we defined solar potential as the amount of power we could generate in 1 year with the optimal solar panel setup, such as [7]:

$$\text{Solar potential (kWh/year)} = N_{\text{panels}} \times \text{Pn} \times \text{PV}_{\text{output}} \qquad (1)$$

where N_{panels} is the maximum number of solar panels that might be installed on the gradient, and Pn (kW) is the supposed capability of a solar plate.

4. Results

We discuss the pipeline's outcomes for every individual part in this section. It is difficult to correctly assess our results without genuine standards to evaluate ourselves against. Calculating the PV output using the position of our roofs, azimuth, and pitch, on the other hand, provided results that are comparable with Gunupur, Odisha's geographical location:

1. When compared to north-facing slopes, south-facing slopes enhance PV production.
2. West-facing slopes have the same PV output as east-facing slopes.

This does not always imply that a south-facing gradient would create more power than another, and it is dependent on the amount of solar panels. Roofs of large structures, in general, increase this output.

We employed the method of verifying the techniques used in each phase independently since we could not discover an open database for our coverage, which would enable us to verify the complete workflow.

Roof section segmentation (with the exception of the backdrop) achieved pixel-wise accuracy of 82%, and roof object segmentation achieved 30%. The substantial performance disparity is accounted for by the discrepancy in training data. Many realistic roof parts were recovered from the city's 3D models, as compared to the tiny number of manually annotated roof objects. This resulted in a significant number of correct roof section labeling as well as pitch and azimuth predictions as shown in Fig. 5.

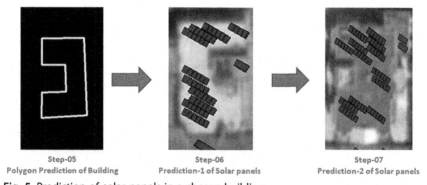

| Step-05 | Step-06 | Step-07 |
| Polygon Prediction of Building | Prediction-1 of Solar panels | Prediction-2 of Solar panels |

Fig. 5 Prediction of solar panels in a chosen building.

Table 1 Ratings for the algorithm using the proposed approach for different tasks.

Task	Model	Score
Roof section segmentation	ResNet-34	Pixel accuracy $= 82\%$
Roof substance segmentation	ResNet-34	Pixel accuracy $= 28\%$
Azimuth	Geometric	Accuracy $= 82\%$
Mean field as a purpose of latitude	Linear regression	$R^2 = 0.93$, MAE $= 4.1°$
Normalize pitch	Random forest	$R^2 = 0.32$, MAE $= 5.8°$

Table 1 shows our approach ratings for algorithms that have a validation set. To forecast the pitch, two separate tasks are employed.

The roofing section pitch regression analysis task score was calculated by matching the arithmetic mean of multiple cities and projecting the final. The final score is calculated by averaging all five combinations and has a relatively high R^2 value of 0.93. This provides a positive association, but the tiny number of latitude values suggests caution when extending these results. The MAE of the normalized pitch model is 0.2, equating to an actual pitch MAE of 5.8. The azimuth model produced good results, with a reliability of 82%. We may consider the effects of pitched and azimuth numbers on a rooftop's final solar potential.

5. Conclusion

In this study, we described a fully functioning pipeline for predicting rooftop solar potential using aerial photography, architectural attributes, and online information labels. The overall 3D geometry design of GIET building in Gunupur is the primary source of ground truth. The process is fully based on structured data and scaled aerial images, which will eventually allow us to anticipate Gunupur's solar energy potential on each and every roof. The fundamental disadvantage of the approach presented in this study is the absence of testing datasets for the whole pipeline, especially for the shading method. As indicated in Section 3, we applied the method of verifying the methods used in each stage individually.

The approach provided here is designed for commercial use, and our primary validation has come from our customers in the energy sector, who have compared our findings to their internal information. Our methodology's two key processes, roof section separation and azimuth prediction, produce excellent results and are well-suited to our current application. Roof segmentation and pitch prediction, on the other hand, produce relatively poor results. The first is due to a scarcity of high-quality labeled data. Its poor results of a pitching prediction step, on either hand, are mostly due to the fact that pitch cannot be reliably calculated from aerial data and has little association with architectural elements.

As previously stated, this limitation is reduced by the pitch's little effect on the ultimate solar potential when compared to other factors such as azimuth and roof section surface. We anticipate that our method will lead to a better understanding of the producing capacity made available by the widespread adoption of solar panel systems on residential and commercial structures, hence hastening the transition to a more sustainable energy system.

References

[1] IEA—International Energy Agency, CO_2 Emissions Statistics, 2018.
[2] IRENA—International Renewable Energy Agency, Renewable Power Generation Costs in 2017, 2018.
[3] Project Google Sunroof. https://www.google.com/get/sunroof.
[4] My Power Engie. https://mypower.engie.fr/.
[5] In Sun We Trust. https://simulateur.insunwetrust.solar/.
[6] RhinoSolar Grand Lyon. https://rhinoterrain.com/fr/rhinosolar.html.
[7] Archelios MAP. https://www.cadastre-solaire.fr/.

CHAPTER FIFTEEN

Experience in using sensitivity analysis and ANN for predicting the reinforced stone columns' bearing capacity sited in soft clays

Tammineni Gnananandarao[a]**, Kennedy C. Onyelowe**[b,c]**,
and K.S.R. Murthy**[d]

[a]Department of Civil Engineering, Aditya College of Engineering and Technology, Surampalem, Andhra Pradesh, India
[b]Department of Civil Engineering, Michael Okpara University of Agriculture, Umudike, Nigeria
[c]Department of Civil Engineering, University of the Peloponnese, Patras, Greece
[d]Adhoc Faculty, Department of Electrical Engineering, National Institute of Technology, Tadepalligudem, Andhra Pradesh, India

Notations

ANFIS	adaptive neuro-fuzzy system
ANNs	artificial neural networks
d/D	ratio of diameter of geogrid layer to diameter of the footing
FIS	fuzzy inference system
GP	genetic programming
L/d_{sc}	ratio of stone column length to diameter of stone column
M5P	M5P model trees
MAE	mean absolute error
MAPE	mean absolute percentage error
MSE	mean squared error
q_{sc}	bearing capacity of the reinforced stone columns sited in soft clays
q_u	bearing capacity of unreinforced stone column
r	correlation coefficient
R2	coefficient of determination
RFR	random forest regression
RMSE	root-mean-square error
s/d	ratio between spacing and diameter of stone column
SVM	support vector machine
t/D	ratio of thickness of sand bed to diameter of footing

1. Introduction

Construction on soft clay soils is quite well known to pose a significant challenging task to geotechnical practitioners. The soft clays are vulnerable in both compression and bearing capacity. Hence, preparation of the soft clay for construction involves modifying its characteristics. Accordingly, geotechnical engineers choose to enhance the ground with stone columns, and it reduces the compression and increases the bearing capacity [1,2]. In the stone columns, geosynthetics were used horizontally as a reinforcing layer [3,4]. Geosynthetic reinforced stone columns on the soft clay further enhances the bearing capacity and reduces the settlement. However, due to the intricate geometry and various uncertainties in the geotechnical parameters, no effective approach for predicting the reinforced stone columns' composite foundation bearing capacity has been presented. In this present study, an ANN technique is taken up to predict the reinforced stone columns' bearing capacity sited in the soft soils.

2. Background

As computing software and hardware systems advance, various computer-aided modeling and soft computing approaches, including as artificial neural networks (ANNs), adaptive neuro-fuzzy, fuzzy inference, M5P models, random forest regression, support vector machine, and genetic programming have been accomplished by a variety of scholars in a variety of civil engineering fields [5,6]. Such computer approaches offer a number of advantages that make them an appealing solution for forecasting various situations. The first distinguishing attribute is that they are data-driven self-adaptive approaches. That is, they do not necessitate numerous previous assumptions about the models of the problem under consideration. They use data to automatically learn the layout of a prediction model. These approaches grow more appealing due to their information processing capabilities, such as nonlinearity, high parallelism, resilience, fault and failure tolerance, and generalizability. Besides, these strategies are being used effectively to tackle difficulties in the civil engineering field.

The purpose of this study is to make use of a powerful branch of artificial neural networks (ANNs), with feed-forward backpropagation algorithm to develop a far more accurate bearing capacity forecasting model of reinforced stone columns sited in soft clays. A broad and reliable bundle of data that include 219 field experimental and laboratory experimental data are gathered in order to create the model. To test the resilience of the resulting model, several performance measure parameters were used.

3. ANNs "artificial neural networks"

ANNs are one of the forms of artificial intelligence (AI) that attempts to imitate the human neural network system. The scope of this study does not allow for a detailed explanation of ANNs. ANNs' construction and functioning have been detailed in detail by several scholars [7–11]. As illustrated in Fig. 1, a typical ANN structure consists of a number of nodes or processing elements, and they usually arranged in layers: input, hidden, and output layer.

3.1 ANN structure selection

Previous research by Boger and Guterman [12] found that the structure of the neural network influences its performance. However, there is no technique that has been established for obtaining the network structure. As a result, the researchers opted to a time-consuming tentative approach. In the case of a multilayer feed-forward network (MFFN), making a choice of number of hidden layers and the number of neurons existing in every hidden layer will always be complicated. Needed to decide on the number of layers and nodes in the hidden layer should be determined. There has been

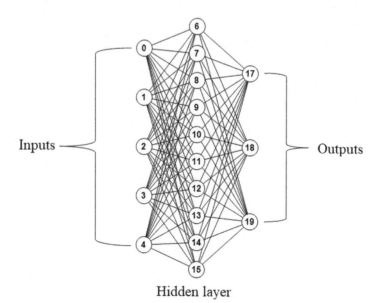

Hidden layer

Fig. 1 Typical structure of ANN model.

no precise formula/procedure for determining the framework of neural networks. Even so, studies have proved that some general guidelines can be followed to begin with. The number of hidden layer neurons could be two-thirds (or ≈70%) of the size of the input layer proposed by Boger and Guterman [12]. According to Boger and Guterman [12], if the number of neurons in the hidden layer is inadequate, the number of neurons in the output layer can be raised later. According to Berry and Linoff [13], the number of neurons in the hidden layer should be decided based on the number of neurons in the input later. Yet, Kurkova [14] claims that the hidden layer neurons will be sized anywhere between the input and output layers. Keeping the preceding in mind, using the thumb rules suggested by Boger and Guterman [12], the number of neurons in the hidden layers was calculated. Other researchers recommend a similar approach [15,16]. In cases where researchers' primary interest is accuracy, multiple hidden layers are used. For the hidden layer, the number of layers and neurons were fixed to decide the finest ANN structure.

The main issue in successfully applying the ANNs model knows when should you stop training? Excessive ANN training leads to noise, but insufficient ANN training leads to bad predictions. As a result, generalization of the network will not result in a new set of data. Hence, for every fixed iteration, the mean squared error (MSE) among the actual and targeted value was calculated and as shown in Fig. 2. For identifying the best ANN structure, the iteration with the lowest MSE is chosen. From Fig. 2, the number

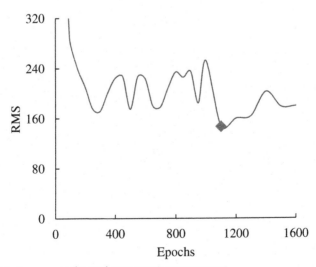

Fig. 2 Plot between epochs and root-mean-square error.

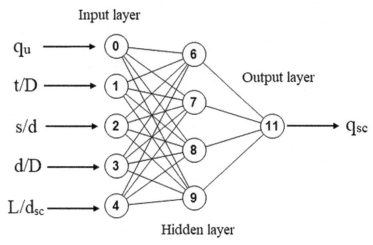

Fig. 3 Architecture of developed ANN model.

iterations were fixed as 1000. Finally, the ANN model selected for this study does have a 5-4-1 structure for way of creating, as shown in Fig. 3.

3.2 Data collection

This present study offering an ANNs application for constitutive modeling of the reinforced stone columns' bearing capacity sited in soft soils.

Table 1 summarizes the experimental data used in this investigation, which include 219 records drawn from field and laboratory studies. The experimental results (data) for training and testing were chosen at random,

Table 1 Variables in the ANN model data ranges.

Parameters (input/output)	Minimum	Maximum	Mean	Stand. dev.
Un-reinforced stone columns' bearing capacity (q_u)	4.21	53.82	38.19	15.72
Ratio of thickness of sand bed to diameter of footing (t/D)	0	0.5	0.22	0.11
Ratio of diameter of geogrid layer to diameter of footing (d/D)	0	4	1.59	1.56
Ratio between spacing and diameter (s/d) of stone column	0.5	20	8.97	6.42
Ratio of stone column length to diameter of stone column (L/d_{sc})	2	8	5.82	1.09
Reinforced stone columns' bearing capacity (q_{sc})	11.39	309.76	133.01	77.35

Table 2 Results of ANN consecutive modeling performance measures.
Performance measures results

For raining data						For testing data					
r	R^2	MSE	RMSE	MAE	MAPE	r	R^2	MSE	RMSE	MAE	MAPE
0.99	0.99	154.55	12.43	9.46	9.67	0.99	0.99	195.08	13.97	10.42	11.23

which is essential to assess the ANN model's generalization capability using the testing data set. Further, the total data were separated into two sets, one is for training and another one is for testing with a percentage of 70 and 30, respectively, as followed by Ref. [7].

3.3 Performance measures

After selection of suitable ANN structure, accuracy is one of the factors used to evaluate the accuracy of a prediction. Therefore, performance measures such as "correlation coefficient" (r), "coefficient of determination" (R^2), "mean square error" (MSE), "root-mean-square error" (RMSE), "mean absolute error" (MAE), "mean absolute percentage error" (MAPE) were used. The detailed explanation about the each above said performance parameter was explained in Refs. [7–11]. The above-mentioned performance measures were calculated and reported in Table 2. From the study of Table 2, it reveals that the statistical performance measures values are within the permissible rage as evident from Ref. [7].

4. Results and discussions

4.1 Model equation and sensitivity analysis

The purpose of this research was to create an equation for accurate prediction of the reinforced stone columns' bearing capacity sited in soft soils. Hence, the Eqs. (1)–(7) were developed using weights and biases gained from the ANN model with the help of sigmoid function and 5-4-1 architecture as reported in Table 3.

$$A = 3.03 - 1.42 \times q_u - 1.12 \times \frac{s}{D} - 0.02 \times \frac{t}{D} - 1.32 \times \frac{d}{D}$$
$$- 1.32 \times \frac{L}{d_{sc}} \tag{1}$$

Table 3 Weights and biases for the ANNs having topology 5-4-1.

	Weights						Biases	
Hidden neurons	q_u (kPa)	s/D (%)	t/D	d/D	L/d_{sc}	q_{rs} (kPa)	b_{hk}	b_0
1	−1.42	−1.12	−0.02	−1.32	−1.32	−2.71	3.03	3.24
2	−0.49	1.22	−2.12	0.49	−0.55	−4.06	1.24	−
3	−0.49	−2.99	−0.58	−1.30	−2.07	−5.14	6.52	−
4	3.31	−2.99	−1.13	2.03	1.21	4.37	4.37	−

$$B = 1.24 - 0.49 \times q_u + 1.22 \times \frac{s}{D} - 2.12 \times \frac{t}{D} + 0.48 \times \frac{d}{D}$$
$$- 0.55 \times \frac{L}{d_{sc}} \tag{2}$$

$$C = 6.52 - 0.49 \times q_u - 2.99 \times \frac{s}{D} - 0.58 \times \frac{t}{D} - 1.30 \times \frac{d}{D}$$
$$- 2.07 \times \frac{L}{d_{sc}} \tag{3}$$

$$D = 4.37 + 3.31 \times q_u - 2.99 \times \frac{s}{D} - 1.13 \times \frac{t}{D} + 2.03 \times \frac{d}{D} 1.21$$
$$\times \frac{L}{d_{sc}} \tag{4}$$

$$E = 3.24 - \frac{2.7}{1 + e^{-A}} - \frac{4.06}{1 + e^{-B}} - \frac{5.14}{1 + e^{-C}} + \frac{4.37}{1 + e^{-D}} \tag{5}$$

$$q_{rs} = \frac{1}{1 + e^{-E}} \tag{6}$$

The q_{rs} value as gained from Eq. (6) is within $[-1, 1]$, and this should be denormalized as

$$q_{rs} \ (\text{kPa}) = 0.5(q_{rs} + 1)\left([q_{rs}]_{max} - [q_{rs}]_{min}\right) + [q_{rs}]_{min} \tag{7}$$

where $[q_{rs}]_{max}$ and $[q_{rs}]_{min}$ are the maximum and the minimum value of predicted bearing capacity of reinforced stone columns sited in soft clays, respectively.

From the literature, it was absorbed that ANN has a capacity to predict the output in precise manner and also it was once again proved that from the present study is that, ANN can predict the desired output precisely as evident from Figs. 4 and 5. The variation of the present data and predicted data was compared to see the difference among them as evident from Figs. 4 and 5. Also, Figs. 4–6 reveal that the proposed equations based on ANNs can

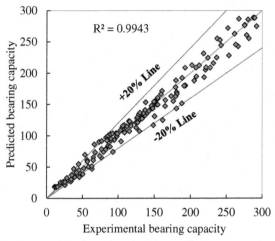

Fig. 4 Predicted vs targeted reinforced stone columns' bearing capacity sited in soft soil straining data set using ANNs.

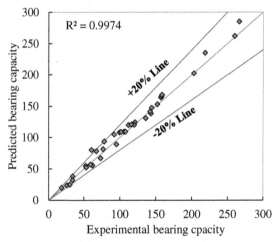

Fig. 5 Predicted vs targeted reinforced stone columns' bearing capacity sited in soft soils testing data set using ANNs.

predict accurately the bearing capacity of reinforced stone columns sited in soft clays.

Finally, sensitivity analysis is conducted to observe the effect of each individual input parameter on the output parameter based on procedure proposed by [17,18]. For every input neuron, this approach calculates the total of the final weights of the connection from input neuron to hidden neurons and the connection from hidden neurons to output. The relative

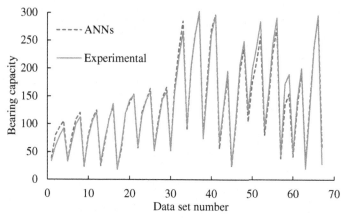

Fig. 6 Data set number vs bearing capacity variation among the experimental data and predicted ANN data.

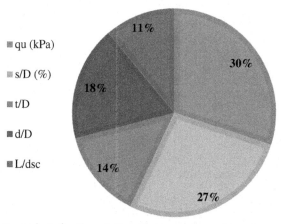

Fig. 7 Sensitivity analysis for the reinforced stone columns' bearing capacity sited in soft clays.

importance of individual input variable on the output of ANN architecture is shown in Fig. 7. From Fig. 7, it concludes that q_u is most influencing parameter with a percentage of 30 as followed by s/D as 27%, d/D as 18%, t/d as 14%, and L/d_{sc} as 11%. The above percentages of the individual input parameter reveal that, each one is having considerable effect on the output; hence, no one is excluded from the ANN model. If the influence of the input parameter is less than 5%, then it can be excluded from the input and the ANN model will be developed from the rest of the input parameters data.

5. Conclusions

Present available proposed theories are the site-specific and don't fit the data in the field. To overcome the above condition, an ANN model equation was developed in the present study. From this study, the drawn conclusions are presented below:

1. The acceptable prediction of the reinforced stone columns' bearing capacity sited in soft clays was achieved with the help of the 5-4-1 architecture.

2. The proposed equations with sigmoid activation function estimates the almost closer values to the experimental bearing capacity values.

3. A mathematical model equation is proposed for predicting the leakage rate which was derived based on the trained weights and biases produced during training with sigmoid activation function.

4. The sensitivity analysis results reveal that, q_u is most influencing parameter on predicting the output (reinforced stone columns' bearing capacity sited in soft soils) with a percentage of 30 as followed by s/D as 27%, d/D as 18%, t/d as 14%, and L/d_{sc} as 11%.

5. The derived neural network model can predict the desired output accurately as evident from the performance measure values.

References

[1] M. Bouassida, B. Jellali, A. Porbaha, Limit analysis of rigidfoundations on floating columns, Int. J. Geomech. 89–101 (2009), https://doi.org/10.1061/(ASCE)1532-3641 (2009)9:3(89).

[2] L. Tang, S. Cong, X. Ling, J. Lu, A. Elgamal, Numerical study on ground improvement for liquefaction mitigation using stone columns encased with geosynthetics, Geotext. Geomembr. 43 (2) (2015) 190–195.

[3] C.S. Wu, Y.S. Hong, The behavior of a laminated reinforced granular column, Geotext. Geomembr. 26 (4) (2008) 302–316.

[4] S. Roy, K. Deb, Bearing capacity of rectangular footings on multilayer geosynthetic-reinforced granular fill over soft soil, Int. J. Geomech. (2017) 04017069, https://doi.org/10.1061/(ASCE)GM.1943-5622.0000959.

[5] A.M. Ebid, 35 years of (AI) in geotechnical engineering: state of the art, Geotech. Geol. Eng. (2020), https://doi.org/10.1007/s10706-020-01536-7.

[6] A.H. El-Bosraty, A.M. Ebid, A.L. Fayed, Estimation of the undrained shear strength of east port-said clay using the genetic programming, Ain Shams Eng. J. (2020), https://doi.org/10.1016/j.asej.2020.02.007.

[7] R.K. Dutta, K. Dutta, S. Jeevanandham, Prediction of deviator stress of sandreinforced with waste plasticstrips using neural network, Int. J. Geosynth. Ground Eng. 1 (2) (2015) 1–12, https://doi.org/10.1007/s40891-015-0013-7.

[8] T. Gnananandarao, R.K. Dutta, V.N. Khatri, Application of artificial neural network to predict the settlement of shallow foundations on cohesionless soils, Geotechn. Appl. 13 (2019) 51–58, https://doi.org/10.1007/978-98113-0368-5_6.

[9] K.C. Onyelowe, T. Gnananandarao, C. Nwa-David, Sensitivity analysis and prediction of erodibility of treated unsaturated soil modified with nanostructured fines of quarry dust using novel artificial neural network, Nanotechnol. Environ. Eng. 6 (2021) 37, https://doi.org/10.1007/s41204-021-00131-2.

[10] R.K. Dutta, R. Rani, T. Gnananandarao, Prediction of ultimate bearing capacity of skirted footingresting on sand using artificial neural networks, J. Soft Comput. Civil Eng. 2 (4) (2018) 34–46, https://doi.org/10.22115/SCCE.2018.133742.1066.

[11] T. Gnananandarao, V.N. Khatri, R.K. Dutta, Bearing capacity and settlement prediction of multi-edge skirted footings resting on sand, Ing. Investig. 40 (3) (2020) 9–21, https://doi.org/10.15446/ing.investig.v40n3.83170.

[12] Z. Boger, H. Guterman, Knowledge extraction from artificial neural network models, IEEE Int. Conf. Comput. Cybern. Simul. 4 (1997) 3030–3035.

[13] Berry M.J.A., Linoff G. (1997) Data Mining Techniques. Wiley, New York. A. Blum, Neural Networks in C++, Wiley, New York, 1992.

[14] V. Kurkova, Kolmogorov's theorem and multilayer neural networks, Neural Netw. 5 (1992) 501–506.

[15] Y. Ito, Approximation capabilities of layered neural networks with sigmoid units on two layers, Neural Comput. 6 (1994) 1233–1243.

[16] W. Sarle, Stopped training and other remedies for overfitting, in: 27th Symposium on the Interface Computing Science and Statistics, Pittsburgh, 1995.

[17] J.D. Olden, D.A. Jackson, Illuminating the "black box": a randomization approach for understanding variable contributions in artificial neural networks, Ecol. Model. 154 (2002) 135–150.

[18] T. Gnananandarao, V.N. Khatri, R.K. Dutta, Prediction of bearing capacity of H shaped skirted footings on sand using soft computing techniques, Arch. Mater. Sci. Eng. 103 (2) (2020) 62–74, https://doi.org/10.5604/01.3001.0014.3356.

CHAPTER SIXTEEN

Sensitivity analysis and estimation of improved unsaturated soil plasticity index using SVM, M5P, and random forest regression

Tammineni Gnananandarao[a], Kennedy C. Onyelowe[b,c], Rakesh Kumar Dutta[d], and Ahmed M. Ebid[e]

[a]Department of Civil Engineering, Aditya College of Engineering and Technology, Surampalem, Andhra Pradesh, India
[b]Department of Civil Engineering, Michael Okpara University of Agriculture, Umudike, Nigeria
[c]Department of Civil and Mechanical Engineering, Kampala International University, Kampala, Uganda
[d]Department of Civil Engineering, National Institute of Technology, Hamirpur, Himachal Pradesh, India
[e]Department of Structure Engineering & Construction Management, Faculty of Engineering, Future University in Egypt, New Cairo, Egypt

Notations

A_c	clay activity
ANFIS	adaptive neuro-fuzzy system
ANNs	artificial neural networks
C_c	clay content
FL	fuzzy logic
GA	genetic algorithm
H_c	hybrid cement percent by weight
LL	liquid limit
M5P	M5P model trees
MAE	mean absolute error
MAPE	mean absolute percentage error
MSE	mean squared error
PI	plasticity index
PL	plastic limit
R	correlation coefficient
R_2	coefficient of determination
RFR	random forest regression
RMSE	root-mean-square error
SVM Poly	support vector mechanics polynomial
SVM RBK	support vector mechanics Radial basis kernel
SVM	support vector machine

1. Introduction

In semiarid regions are dry but slightly having rains as an example India. In such regions, swelling and shrinkage cycles can be observed in the soils due to the seasonal changes. To avoid such swell and shrink cycles, many techniques are adopted to stabilize the soils [1–5]. In the recent past, stabilization of soil using nanostructured materials came into the light. Unsaturated soils treated with nanostructured materials have improved shear and compressive strength properties. Nanostructured materials can also improve soil consistency, which refers to the strength with which the soil components are kept together or their resistance to deformation and rupture. Hence, to generate the stabilized soil data, expensive laboratory or in situ tests needs to be conducted. To overcome the above extensive tests, soft computing techniques may be used to get the prediction of desired output from the preexisting data [6].

2. Background

In the recent times, significant evolution that was observed in the global optimization inspired by soft computing techniques and artificial intelligence has developed [7–11]. Due to their great optimization capability in different problems, they have been used in the civil engineering applications for the better prediction of the nonlinear data in the near past. Strategies of soft computing are having many advantages over the existing methods (having some limitations and assumptions) for use in civil engineering and especially in geotechnical engineering for development and practice [12,13].

Literature has been shown that soft computing techniques and artificial intelligence can advance the efficiency of prediction of nonlinear data [10]. Taken as a whole, the fields of statistics, numerics, and optimizing techniques present critical issues, which help to describe the dataset properties, discover relations and patterns within the dataset, and select and implement the appropriate model using artificial intelligence. Many evolutionary soft computation techniques were used to estimate engineering factors, for example "fuzzy logic" (FL), M5P model tree, "random forest regression" (RFR), "artificial neural networks" (ANNs), "support vector machine" (SVM), "genetic programming" (GP), "genetic algorithm" (GA), "evolutionary polynomial regression" (EPR), and "adaptive neuro-fuzzy inference system" (ANFIS) [14–25]. Each soft computing technique is having unique versatile robust prediction power for predicting the desired output parameter. These AI techniques have been successfully applied in estimating the (USC) of normal and high-performance concretes [14] shear capacity of FRP concrete beams [15], "ultimate bearing capacity" (qu) of strip foundation rested on dense sand [16], on loose sand deposit [17], on rock mass [18], near slopes [19], cone resistance in cohesive soils [20,21], elastic modulus for granite rock [22], shrinkage limit, compressibility index and (UCS) of nanocomposite binder-treated soils [23–25], respectively. In the current study, SVM, M5P, and RFR models were used for the prediction of the unsaturated soil plasticity, with the help of (HC) % by weight (Cc), (Ac), (LL), and (PL).

3. Soft computing techniques

3.1 "Support vector machines" (SVM)

SVM was invented by [26] with an alternate ε-insensitive loss function. It allows for regression problems to use the definition of margin. The boundary is identified as the total distances of hyper-planes to the nearest point of the two categories. The main aim of SVM is to figure out the function with maximum deviation (ε) from the real desire vectors considering all provided training dataset, besides that, it must be flat function as possible [27], where a kernel function concept was introduced by Vapnik [26] for nonlinear SVM regression. The enthusiastic readers are advised to refer for more descriptions of supporting vector regression [26,27].

3.1.1 Details of kernel

In SVM, a kernel function concept was used, where the nonlinear decision surface circumstances occurred [26]. Many numbers of "kernel functions" are introduced in the past decade, but the literature [28–30] indicated that polynomial kernel and radial kernel perform better for geotechnical engineering applications. Hence, in the present article, polynomial kernel $K(x,y) = (x \cdot y)^d$ and radial kernel $e^{-\gamma|x-y|2}$ were used (d,γ are the kernel parameters). In order to use SVM, suitable user-defined parameters have to be set first. These used-defined parameters are playing a major role in SVM prediction. The SVM needs kernel-specific parameters in addition to the selection of a kernel. The appropriate values of "regulatory parameter" (C) and "size of the error-insensitive zone" (ε) should be evaluated. A manual procedure was followed to select the values of (C, γ, and d), which involves performing a series of trials by means of different combinations of C and d for the polynomial kernel; C and γ for the RBF kernel support vector machines (SVMs). Correspondingly, several trials were conducted in order to find the appropriate value for ε the error-insensitive zone having a constant value of C and defined kernel parameters. The value of $C = 0.011$ is found to be good for this study. In this article, radial function kernel and the polynomial kernel of the (SVM) are represented as SVM_{RBFK} and SVM_{POLYK}, respectively.

3.2 M5P model tree

MTs, in spite of the fact that straightforward, are proficient and exact technique for simulating the pattern and relations of huge datasets [31]. Quinlan et al. [26] created a modern sort of tree called M5 to estimate continuous variables. An overfitting issue can happen amid MT development according to training data. Predictably, the precision of the tree for training cases raises monotonically as the tree develops. Anyways, this raises overfitting; in this way, the precision measured over the independent test cases to begin with raising, then decreases. A strategy for minimizing this issue is called "pruning." Finally, the sharp discontinuities occurred at the intersections between adjacent linear models are smoothed using leaf technique after Quinlan et al. [32]. The predicted value is filtered and smoothed node-by-node back to the root by integrating with the linear model value at each node. In short, M5 tree technique has three main steps: (a) constructing the tree; (b) pruning the tree; and (c) smoothing the tree. Construction of the M5 tree aims to maximize a term called "standard deviation reduction" (SDR) which is:

$$\text{SDR} = sd(T) - \sum_i \frac{|T_i|}{|T|} \times sd(T_i)$$

where,

T: number of cases,

T_i: ith subset of cases which produced by dividing the tree according to a set of variables

$sd(T)$: standard deviation of T,

$sd(T_i)$: standard deviation of T_i, as a measurement of error [26].

Wang et al. [33] updated the first M5 tree to handle listed attributes and property lost attributes; they named the updated algorithm "the M5P algorithm." Within the M5P tree calculation, all identified properties are changed into binary factors before constructing the tree. The fundamental tree is shaped utilizing the dividing criterion that treats the "standard deviation of the class values" which submitted to a node as error measurement, and computes the estimated error decreasing as a function of examine each property at the considered node. The property that maximizes the estimated error decreasing is elected. M5P MT technique is not widely used in the water studies yet [34].

3.3 "Random forest regression" (RF) regression

RF is a technique used to classify data and predict information. It combining predicting trees, each one is developed based on random data subset selected from the total input dataset. For regression, the predicting trees take numerical values instead of "class labels" utilized for classification [35]. RF regression was utilized current research to handle the randomly chosen variables at each considered node of the developed tree. Bagging is a strategy to create a training dataset by arbitrarily drawing considering substitution N samples; N is "size of the original training set" [36], or an arbitrarily chosen portion of the training database that will be utilized for the development of individual trees for each feature. For bagging "bootstrap test," training dataset comprises around 67% of the total dataset, in this way almost 33 and of the dataset is cleared out from each developed tree. These cleared out dataset are called "out-of-bag" or "out of the bootstrap testing." Considering criteria for variable selection and pruning technique are basic requirements to generate a prediction tree. A number of approaches to determine variables for prediction tree are proposed in previous researches. Most approaches proposed direct correlation between quality measurement and variables. The used criteria for variable selection are "information gain ratio criterion" [37] &"Gini Index" [38]. RF regression approach used in this research utilized "Gini Index" as variable selection criteria that estimate the "impurity" of a variable with regard to the outcomes. The configurations of RF regression permit a tree to develop to the maximum boundaries of new training dataset employing a combination of factors. The full-grown prediction trees aren't pruned back. It is one of the main points of interest of RF regression over other tree techniques [37]. Previous research's indicated that the selection of the pruning strategies is the main factor affected the behavior of the prediction tree not the variable choosing criteria [39]. Breiman [35] illustrated that increasing the number of trees converges the generalized error regardless of the tree pruning, and also the overfitting isn't an issue due to "Strong Law of Large Numbers" [40]. Number of factors utilized for each node to create a tree and the number of trees to be developed are two parameters determined by user to conduct RF regression [35]. Only the selected parameters are considered for the optimum division for each node. For RF regression, the outcome values are numerical, hence the MSE can be calculated. The RF predictor is computed as the average of the errors from all developed trees.

3.4 Data collection

SVM, M5P, and random forest regression will be used for predicting the improved unsaturated soil plasticity index of stabilized soil. For preparing the SVM, M5P, and random forest regression models, data that were collected from the literature are presented in the Table 1 in the form of statistical analysis for inputs and output.

The collected database was separated into 2 sets, training set and testing set. For training, 70% of data were selected and rest of the data (30%) are for testing based on the [16,22,41]. SVM, M5P, and random forest regression algorithms were applied in both training and testing phases to estimate the output. To see the accuracy of the predicted model corresponding to the used soft computing technique (SVM, M5P, and random forest regression), performance measures were used.

3.5 Performance measures

Performance measures (statistical parameters) were utilized to monitor the accuracy of the proposed soft computing strategies such as SVM, M5P, and random forest regression. Performance measures were considered for the present study are "correlation coefficient" (r), "coefficient of determination" (R^2), "mean-squared error" (MSE), "root-mean-square error" (RMSE), "mean absolute error" (MAE), and "mean absolute percentage error" (MAPE). To validate the mode by using the above said performance measures, it needs the range of each parameter. Hence, if the r and R^2 close to 1 and MSE, RMSE, and MAE are close to zero then it is a perfect model

Table 1 Ranges of used variables for ANN model.

Variable (input/output)	Min.	Max.	Mean	Standard deviation
(HC)	0	0.12	0.06	0.035
(CC)	0.23	0.24	0.24	0.003
(A_c)	0.6	2	1.3	0.39
(LL)	0.27	0.66	0.48	0.12
(PL)	0.13	0.21	0.17	0.02
(I_p)	0.14	0.45	0.31	0.09

[42,43]. It is pertinent to mention here that each performance measures mathematical expression is provided in the literature [8].

3.6 Analysis results and detailed discussions

The goal of the current research is to predict the unsaturated soil plasticity index using SVM, M5P, and RFR. The randomly selected data for training and testing were utilized in the SVM, M5P, and RFR and the obtained prediction results of training as well as testing were validated with the help of the performance measures for each model and the results are presented in Figs. 1 and 2, respectively. From the study of Figs. 1 and 2, it reveals that among the four models, SVM RBK is more accurate than the remaining SVM poly,

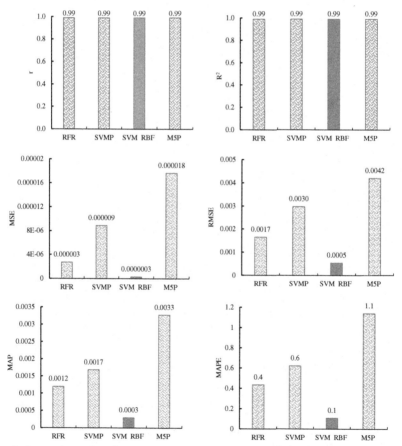

Fig. 1 Comparison of performance measures among the different soft computing techniques for training data.

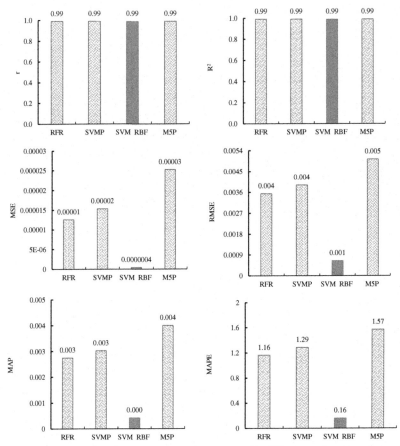

Fig. 2 Comparison of performance measures among the different soft computing techniques for testing data.

M5P, and RFR to predict the unsaturated soil plasticity index based on the range of r, R^2, MSE, RMSE, MAP, and MAPE. Further, the predicted vs actual values of unsaturated soil plasticity index plots were depicted in Figs. 3–6. From the study of Figs. 3–6, it reveals that the SVM RBK model is predicting more accurately than the other models as evident from the overall performance measures comparison. Hence, in the proceeding section, the optimum model (SVM RBK) was implemented to perform the sensitivity study.

3.7 Sensitivity analysis

After completion of model, the best predicted method was chosen for the sensitivity analysis. Sensitivity analysis will give the importance of the input

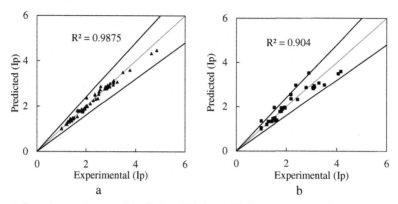

Fig. 3 Experimental vs predicted plot of RFR model for training (A) and testing (B).

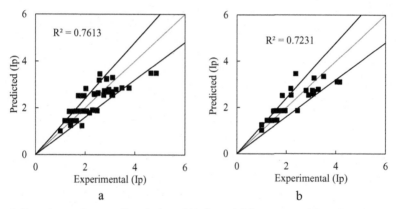

Fig. 4 Experimental vs predicted plot of M5P model for training (A) and testing (B).

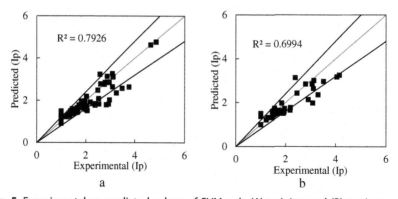

Fig. 5 Experimental vs predicted values of SVM poly (A) training and (B) testing.

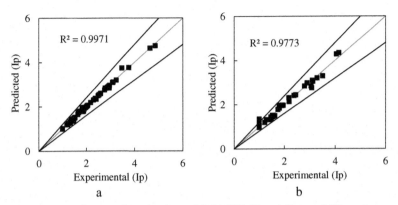

Fig. 6 Experimental vs predicted values of SVM RBK (A) training and (B) testing.

Table 2 Sensitivity analysis for the SVM RBK kernel model.

Input combinations	Input parameter removed	SVM RBK					
		r	R^2	MSE	RMSE	MAE	MAPE
H_c, C_c, A_c, LL, PL	–	0.99	0.99	0.0000004	0.0007	0.0004	0.1358
C_c, A_c, LL, PL	H_c	0.99	0.99	0.0000005	0.0007	0.0004	0.1239
H_c, A_c, LL, PL	C_c	0.99	0.99	0.0000003	0.0005	0.0003	0.0790
H_c, C_c, LL, PL	A_c	0.99	0.99	0.0000006	0.0008	0.0004	0.1788
H_c, C_c, A_c, PL	LL	0.99	0.99	0.0000019	0.0014	0.0008	0.2949
H_c, C_c, A_c, LL	PL	0.99	0.99	0.0000025	0.0016	0.0009	0.3148

parameter to predict the output parameter. To achieve the above, first all the input parameters were considered for the prediction of the desired output and corresponding performance were calculated. Further, same procedure was followed by removing the single input parameter from the total input parameters. Finally, the calculated performance measure values for each remove of the input parameter were present in Table 2. From the study of Table 2, it reveals that the most influencing parameter is PL followed by the LL, A_c, H_c, and C_c.

4. Conclusions

This research has studded the capacities of the computational methods for sensitivity analysis and prediction improved unsaturated soil plasticity index of stabilized soil. From the above study, finally drawn conclusions were presented below:

1. The predicted results are fairly good and recommend the application of SVM RBK, SVM poly, M5P, and RFR for analyzing the improved unsaturated soil plasticity index of stabilized soil.

2. Comparing the behavior of the evaluation parameters indicated that RF technique is more successful than M5 tree and ANN for the considered dataset.

3. The results of sensitivity study indicate that plastic limit (PL) is the most important parameter for measuring the unsaturated soil plasticity index of stabilized soil.

References

[1] A.A. Abiodun, Z. Nalbantoglu, Lime pile techniques for the improvement of clay soils, Can. Geotech. J. 52 (6) (2015) 760–768.

[2] P. Ashok, G.S. Reddy, Lime pile technique for the improvement of properties of clay soil, Int. J. Sci. Res. 5 (11) (2016) 1204–1210.

[3] F. Darikandeh, B.V.S. Viswanadham, Swell behavior of expansive soils with stabilized fly ash columns, in: Indian Geotechnical Conference, Chennai, India, 2016.

[4] A.K. Jha, P.V. Sivapullaiah, Lime stabilization of soil: a physico-chemical and micro-mechanistic perspective, Indian Geotech. J. 50 (3) (2019) 339–347.

[5] S.Y.S. Karthick, R. Vasanthanarayanan, S. Ayswarya, C. Meenakshi, Soil stabilization using plastics and gypsum, Int. J. Res. Eng. Sci. Manage. 2 (1) (2019) 351–360.

[6] A.M. Ebid, L.I. Nwobia, K.C. Onyelowe, F.I. Aneke, Predicting nanobinder-improved unsaturated soil consistency limits using genetic programming and artificial neural networks, Appl. Comput. Intell. Soft Comput. 2021 (Article ID 5992628) (2021) 1–13, https://doi.org/10.1155/2021/5992628.

[7] K.C. Onyelowe, J. Shakeri, Intelligent prediction of coefficients of curvature and uniformity of hybrid cement modified unsaturated soil with NQF inclusion, Clean. Eng. Technol. 4 (2021).

[8] T. Gnananandarao, R.K. Dutta, V.N. Khatri, Application of artificial neural network to predict the settlement of shallow foundations on cohesionless soils, in: Indian Geotechnical Conference, 15–17 December, IIT Madras, Chennai, India, 2016.

[9] K.C. Onyelowe, T. Gnananandarao, C. Nwa-David, Sensitivity analysis and prediction of erodibility of treated unsaturated soil modified with nanostructured fines of quarry dust using novel artificial neural network, Nanotechnol. Environ. Eng. 6 (37) (2021) 1–11.

[10] T. Gnananandarao, R.K. Dutta, V.N. Khatri, Bearing capacity and settlement prediction of multi-edge skirted footings resting on sand, Ing. Invest. J. 40 (3) (2020) 9–21.

[11] T. Gnananandarao, R.K. Dutta, V.N. Khatri, Application of artificial neural network to predict the settlement of shallow foundations on cohesionless soils, Geotechn. Appl. Lect. Notes Civil Eng. 13 (2019) 51–58, https://doi.org/10.1007/978-981-13-0368-5_6.

[12] Y. Singh, P. Bhatia, O. Sangwan, A review of studies on machine learning techniques, Int. J. Comput. Sci. Secur. 1 (2007).

[13] A.M. Ebid, 35 years of (AI) in geotechnical engineering: state of the art, Geotech. Geol. Eng. (2020), https://doi.org/10.1007/s10706-020-01536-7.

[14] A. Behnood, V. Behnood, M.M. Gharehveran, K.E. Alyamac, Prediction of the compressive strength of normal and high-performance concretes using M5P model tree algorithm, Constr. Build. Mater. 142 (2017) 199–207.

[15] A.M. Ebid, A. Deifalla, Prediction of shear strength of FRP reinforced beams with and without stirrups using (GP) technique, Ain Shams Eng. J. (2021), https://doi.org/10.1016/j.asej.2021.02.006.

[16] D. Padmini, K. Ilamparuthi, K.P. Sudheem, Ultimate bearing capacity prediction of shallow foundations on cohesionless soils using neurofuzzy models, Comput. Geotech. 35 (2008) 33–46.

[17] R.K. Dutta, T. Gnananandarao, A. Sharma, Application of random forest regression in the prediction of ultimate bearing capacity of strip footing resting on dense sand overlying loose sand deposit, J. Soft Comput. Civil Eng. 2–3 (2018) 01–11.

[18] S. Tajeri, E. Sadrossadat, J.B. Bazaz, Indirect estimation of the ultimate bearing capacity of shallow foundations resting on rock masses, Int. J. Rock Mech. Mining Sci. 80 (2015) 107–117.

[19] A.M. Ebid, K.C. Onyelowe, E.E. Arinze, Estimating the ultimate bearing capacity for strip footing near and within slopes using AI (GP, ANN, and EPR) techniques, J. Eng. (2021), https://doi.org/10.1155/2021/3267018. 11 p., Article ID 3267018.

[20] T. Gnananandarao, R.K. Dutta, V.N. Khatri, Neural networks based prediction of cone side resistance for cohesive soils, Lecture Note Civil Eng. 137 (2021) 389–399.

[21] A.H. El-Bosraty, A.M. Ebid, A.L. Fayed, Estimation of the undrained shear strength of east port-said clay using the genetic programming, Ain Shams Eng. J. (2020), https://doi.org/10.1016/j.asej.2020.02.007.

[22] T. Gnananandarao, R.K. Dutta, V.N. Khatri, Support vector machinesbased prediction of elastic modulus for granite rock, Recent Adv. Comput. Exp. Mech. 1 (2022) (in press).

[23] K.C. Onyelowe, A.M. Ebid, L.I. Nwobia, I.I. Obianyo, Shrinkage limit multi-ai-based predictive models for sustainable utilization of activated rice husk ash for treating expansive pavement subgrade, Transp. Infrastruct. Geotechnol. (2021), https://doi.org/10.1007/s40515-021-00199-y.

[24] K.C. Onyelowe, A.M. Ebid, L. Nwobia, et al., Prediction and performance analysis of compression index of multiple-binder-treated soil by genetic programming approach, Nanotechnol. Environ. Eng. 6 (2021) 28, https://doi.org/10.1007/s41204-021-00123-2.

[25] K.C. Onyelowe, A.M. Ebid, M.E. Onyia, L.I. Nwobia, Predicting nanocomposite binder improved unsaturated soil UCS using genetic programming, Nanotechnol. Environ. Eng. 6 (2021) 39, https://doi.org/10.1007/s41204-021-00134-z.

[26] V.N. Vapnik, The Nature of Statistical Learning Theory, Springer-Verlag, New York, 1995.

[27] A.J. Smola, Regression estimation with support vector learning machines, 1996 (Master's thesis)., Technische Universitat München, Germany.

[28] M. Pal, P.M. Mather, Support Vector Classifiers for Land Cover Classification, January 28–31, Map India, New Delhi, 2003.

[29] M. Pal, A. Goel, Prediction of the end depth ratio and discharge in semicircular and circular shaped channels using support vector machines, Flow Meas. Instrum. 17 (2006) 50–57.

[30] M.K. Gill, T. Asefa, M.W. Kemblowski, M. Makee, Soil moisture prediction using support vector machines, J. Am. Water Resour. Assoc. 42 (2006) 1033–1046.

[31] M.R. Nikoo, A. Karimi, R. Kerachian, H. Poorsepahy-Samian, F. Daneshmand, Rules for optimal operation of reservoir-river-groundwatersystemsconsidering waterqualitytargets: Applicationof M5Pmodel.Water Resour, Manage 27 (2013) 2771–2784.

[32] J.R. Quinlan, Learning with continuous classes, in: Adams, Sterling (Eds.), Proceedings AI'92, World Scientific, 1992, pp. 343–348.

[33] Y. Wang, I.H. Witten, Inducing model trees for continuous classes, in: M. van Someren, G. Widmer (Eds.), Proceedings of the Poster Papers of the 9th European Conference on Machine Learning (ECML 97), 1997, pp. 128–137.

[34] S.N. Almasi, R. Bagherpour, R. Mikaeil, Y. Ozcelik, H. Kalhori, Predicting the building stone cutting rate based on rock properties and device pullback amperage in quarries using M5P model tree, Geotech. Geol. Eng. 35 (2017) 1311–1326.

[35] L. Breiman, Random Forests—Random Features. Technical Report 567, Statistics Department, University of California, Berkeley, 1999. ftp://ftp.stat.berkeley.edu/pub/users/breiman.

[36] L. Breiman, Bagging predictors, Mach Learn. 24 (2) (1996) 123–140.

[37] J.R. Quinlan, Learning with continuous classes, in: Proceedings of Australian Joint Conference on Artificial Intelligence, World Scientific Press, Singapore, 1992, pp. 343–348.

[38] L. Breiman, J.H. Friedman, R.A. Olshen, C.J. Stone, Classification and Regression Trees, Wadsworth, Monterey, 1984.

[39] M. Pal, P.M. Mather, An assessment of the effectiveness of decision tree methods for land cover classification, Remote Sens. Environ. 86 (4) (2003) 554–565.

[40] W. Feller, An Introduction to Probability Theory and its Application, third ed., vol. 1, Wiley, New York, 1968.

[41] B. Singh, P. Sihag, K. Singh, Modelling of impact of water quality on infiltration rate of soil by random forest regression, Model. Earth Syst. Environ. 3 (2017) 999–1004, https://doi.org/10.1007/s40808-017-0347-3.

[42] H. Xu, J. Zhou, P.G. Asteris, D.J. Armaghani, M.M. Tahir, Supervised machine learning techniques to the pre-diction of tunnel boring machine penetration rate, Appl. Sci. 9 (2019) 1–19, https://doi.org/10.3390/app9183715.

[43] H. Harandizadeh, D.J. Armaghani, M. Khari, A new development of ANFIS-GMDH optimized by PSO to predict pile bearing capacity based on experimental datasets, Eng. Comput. 1-16 (2019), https://doi.org/10.1007/s00366-019-00849-3.

CHAPTER SEVENTEEN

Forecasting off-grid solar power generation using case-based reasoning algorithm for a small-scale system

Aadyasha Patel[a] and O.V. Gnana Swathika[b]
[a]School of Electrical Engineering, Vellore Institute of Technology, Chennai, India
[b]Centre for Smart Grid Technologies, School of Electrical Engineering, Vellore Institute of Technology, Chennai, India

Abbreviations and nomenclature

A	Ampere
AI	artificial intelligence
A_t	actual value
CBR	case-based reasoning
F_t	forecasted value
MAD	mean absolute deviation
MAPE	mean absolute percentage error
ML	machine learning
MSE	mean square error
PLX-DAQ	parallax microcontroller data acquisition
RMSE	root-mean-square error
V	volt
W	watt
n	number of values
p, q	Euclidean space points
q_i, p_i	Euclidean vectors

1. Overview

The solar power generation is in the spotlight since it is eco-friendly in nature and offers economic benefits. The prediction of the generated energy has become a part of the solar power generation system. All the countries are nowadays encouraging research, development and contributing toward this nonpolluting system. There are still some challenges associated with solar photovoltaics. Its output is highly ambiguous and varied as a result of dependence on the environmental factors such as atmospheric aerosol level, speed of the wind, irradiance, expanse of cloud cover, temperature, humidity, and settlement of dust particles. Power prediction can lift some of these problems of solar photovoltaics by implementing an improved version of power prediction algorithms.

2. Literature review

Machine learning generally confronts a particular problem of missing/erroneous data. To avoid this, multiple prediction algorithms are available to be used to correct the data. CBR, as explained by Richter and Aamodt [1], Aamodt and Plaza [2], Watson and Marir [3], and Kolodner [4], is one such type of algorithm. It takes references out of the past experiences to fetch solutions for the present problems. For instance, a mechanic, a doctor, a programmer, and a lawyer find the solution suitable to the present problems by experience and knowledge gained over the past. The study of Bjurén [5] shows that CBR is also used for forecasting data for a short time period for energy consumption and usage. Short-term predictions are made available for homes to analyze and consume power reasonably. The CBR uses a crisp data set, which when replaced by fuzzy data set used by Somi et al. [6] and Lu et al. [7], the predicted results become more accurate and clear. CBR is also capable of handling erroneous or missing data. The recorded parameters may have a few data missing or may have recorded wrong value. Articles of Löw et al. [8], Seo et al. [9], and Xie et al. [10] resolves this absent or flawed data problem by implementing CBR algorithm. The medical field also uses this algorithm widely as reported by Recio-García et al. [11] and Oyelade and Ezugwu [12] to check on the medical history of patients for described illnesses. Various industries like chemical and petroleum also make use of CBR for making predictions and checking on missing values as stated by Shokouhi et al. [13] and Zhao et al. [14]. The discussions of Patel et al. [15], Lieber et al. [16], Henriet et al. [17], Knight et al. [18] are briefly about the interpolation method of CBR for prediction of missing values. The power and accuracy of interpolation are essentially investigated. The authors [19–23] have summarized in detail the process of CBR in detail for multiple applications in multiple fields.

3. AI/ML in forecasting and imputation of missing values

In this chapter, CBR algorithm is implemented for forecasting and imputation of missing values. The algorithm is applied in two ways as follows:

CBR without interpolation.

CBR with interpolation.

3.1 CBR algorithm

CBR algorithm is principally implemented when a bulky data pool is available but the concepts elucidating the circumstances are unidentified. It is a kind of a problem-solving technique, which takes references out of the past experiences and fetches solutions for the present problems. This technique is said to be inspired by the day-to-day decisions human beings make. For instance, a mechanic, a doctor, a programmer, and a lawyer find the solution suitable to the present problems by experience and knowledge gained over the past. The whole process of CBR is showed in Fig. 1.

As a computer reasoning tool, CBR comprises of four steps as follows:

Retrieve—fetches the past instances from the previous instance pool of data.

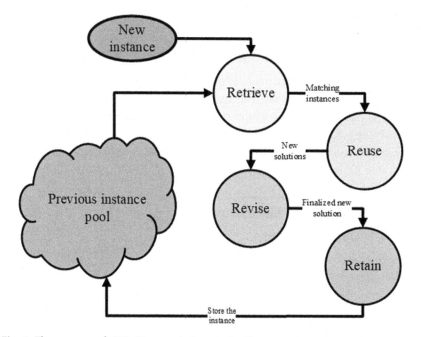

Fig. 1 The process of CBR. *(No permission required.)*

Reuse—matches the current and past instances and also makes adjustments to the situation if needed.

Revise—implements the new solution.

Retain—saves the applied solution for another future instance to the previous instance pool of data.

The CBR algorithm learns from its past instances, it is spontaneous and instinctual, overall understanding of all the instances is not required rather knowledge of significant characteristics is necessary. CBR needs huge amount of space for storage of all the past instances, it may take a longer period of time to trace an equivalent instance from the previous instance pool of data, and the instances may not follow the suitable standards. The CBR algorithm provides solutions on a pretty much approximate scale.

3.2 Euclidean distance equation

The shortest distance between any two points is always a straight line. Euclidean distance equation finds that shortest possible distance using Eq. (1). All the past instances are considered, and the best match is found for the current problem. This method is the simplest and most commonly used to find the disparity in the recorded data.

$$d(p, q) = \sqrt{\sum_{i=1}^{n} (q_i - p_i)^2} \tag{1}$$

3.3 Interpolation technique

Mathematically, interpolation means building a new data set formulated from the discrete recorded data set. The Euclidean distance equation is first used to create the new data model. Then, the actual recorded data are interpolated. The missing values present in the recorded data set are compared with the new created data model. This way the erroneous/missing values are filled in.

4. Results and discussion

4.1 Data collection

For collection of data, open-source hardware and software are chosen since they cost lesser. Arduino Uno is particularly chosen as it provides much flexibility, costs less, and is easily available in the markets. It has a widespread developer community, which provides assistance when required through various blogs and websites. To generate electricity, a 10 W solar panel is

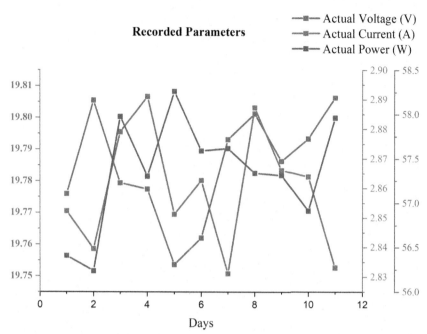

Fig. 2 The recorded parameters. *(No permission required.)*

installed to the roof-top connected with sensors to sense the electrical parameters. Data are instantly logged to the PLX-DAQ excel spreadsheet, which is an add-on provided for MS Excel. The data are logged with time stamp between 12 noon and 2 PM with a gap of 5 s between each consecutive entry. These data are analyzed using CBR algorithm executed in two different ways, i.e., implementing the algorithm with and without interpolation. Due to some fault of the sensors or any other component, an erroneous data may be logged or a data might go missing. To correct this, data prediction using CBR is employed. Fig. 2 illustrates the electrical parameters (voltage, current, and power) that is recorded. The green colored y-axis on the left represents the average of recorded voltages for each day. The red and blue colored y-axes on the right represent the average of recorded currents and powers for each day, respectively.

4.2 Error metrics

The analysis and investigation of the capability of CBR executed with and without interpolation to predict the erroneous/missing data of an off-grid solar power small-scale system is carried out using the error metrics from Table 1.

Table 1 Error metrics.

S. no.	Error metrics
1	$\text{MAD} = \dfrac{\sum_{t=1}^{n} |A_t - F_t|}{n}$
2	$\text{MSE} = \dfrac{\sum_{t=1}^{n} (A_t - F_t)^2}{n}$
3	$\text{RMSE} = \sqrt{\dfrac{\sum_{t=1}^{n} (A_t - F_t)^2}{n}}$
4	$\text{MAPE} = \dfrac{\sum_{t=1}^{n} \left|\frac{A_t - F_t}{A_t}\right|}{n}$

4.3 Comparison of forecasted values using CBR with and without interpolation

The comparison of the values predicted using CBR with and without interpolation is carried out. The analysis is showed in Figs. 3–5. The comparison between voltages is depicted in Fig. 3. The green colored y-axis on the left represents the average of actual recorded voltages for each day. The red and blue colored y-axes on the right represent the average of forecasted voltages using CBR without and with interpolation for each day, respectively. The comparison between currents is depicted in Fig. 4. The green colored y-axis on the left represents the average of actual recorded currents for each day.

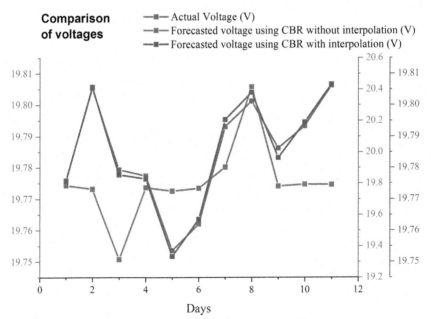

Fig. 3 Comparison of voltages. *(No permission required.)*

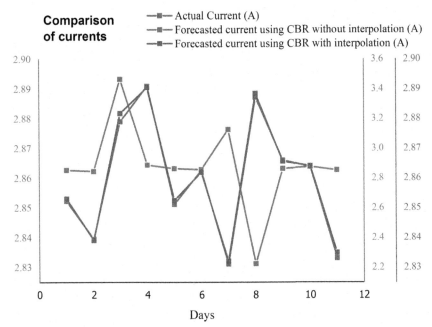

Fig. 4 Comparison of currents. *(No permission required.)*

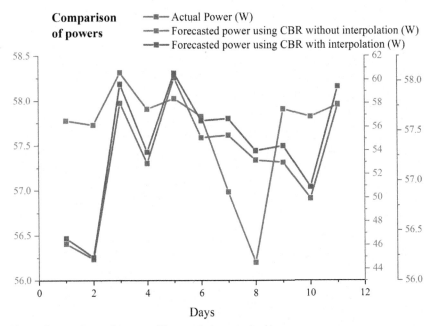

Fig. 5 Comparison of power. *(No permission required.)*

Table 2 Comparison of error metrics.

	Forecasting using CBR							
	Without interpolation				With interpolation			
Parameters	MAD	MSE	RMSE	% MAPE	MAD	MSE	RMSE	% MAPE
Voltage	0.42	0.259	0.508	2.121	0.367	0.198	0.444	1.856
Current	0.394	0.26	0.499	14.078	0.366	0.199	0.445	13.073
Power	8.714	115.091	10.685	15.486	7.85	90.722	9.494	13.979

The red and blue colored y-axes on the right represent the average of forecasted currents using CBR without and with interpolation for each day, respectively. The comparison between powers is depicted in Fig. 5. The green colored y-axis on the left represents the average of actual recorded powers for each day. The red and blue colored y-axes on the right represent the average of forecasted powers using CBR without and with interpolation for each day, respectively. The plots also show that the forecasted parameters obtained by including interpolation technique and the actually recorded values are approximately same. Hence, the interpolation technique gives better performance than the method implemented without interpolation.

The error metrics mentioned in Table 1 are computed and contrasted. The result obtained is displayed in Table 2. It presents the quantitative outcomes obtained from the error metrics. From Table 2, it is understood that implementing CBR with interpolation gives accurate and precise predictions of missing data compared with the predictions obtained using CBR without interpolation.

From Table 2, it is deduced that the MAD reduces by 9.921%, MSE by 21.174%, and RMSE by 11.138% and the %MAPE has a difference of 1.507%. All these parameters prove that the interpolation technique gives much better performance and is outright capable of forecasting accurate values.

5. Applications of CBR in forecasting and imputation of missing values

CBR is nowadays applied to almost all the fields for filling in the missing values like in the engineering domain, medical domain, and business domain. But it has not yet found a place in the academic domain.

6. Summary

The work done here is a comparison done between the forecasts made by CBR algorithm implemented using two ways, i.e., with and without interpolation. The actual and forecasted values by both the techniques are examined and compared. The results show that interpolation technique performs much better than the other technique.

References

[1] M.M. Richter, A. Aamodt, Case-based reasoning foundations. Knowl. Eng. Rev. 20 (3) (2005) 203–207, https://doi.org/10.1017/S0269888906000695.

[2] A. Aamodt, E. Plaza, Case-based reasoning: Foundational issues, methodological variations, and system approaches, Artif. Intell. Commun. 7 (1) (1996) 39–59. [Online]. Available: https://ibug.doc.ic.ac.uk/media/uploads/documents/courses/CBR-AamodtPlaza.pdf.

[3] I. Watson, F. Marir, Case-based reasoning: a review. Knowl. Eng. Rev. 9 (4) (1994) 327–354, https://doi.org/10.1017/S0269888900007098.

[4] J.L. Kolodner, An introduction to case-based reasoning. Artif. Intell. Rev. 6 (1) (1992) 3–34, https://doi.org/10.1007/BF00155578.

[5] J. Bjurén, Using Case-Based Reasoning for Predicting Energy Usage, 2013.

[6] S. Somi, N. Gerami Seresht, A.R., Fayek, Framework for risk identification of renewable energy projects using fuzzy case-based reasoning. Sustainability 12 (13) (2020), https://doi.org/10.3390/su12135231.

[7] J. Lu, D. Bai, N. Zhang, T. Yu, X. Zhang, Fuzzy case-based reasoning system, Appl. Sci. 6 (7) (2016) 189, https://doi.org/10.3390/app6070189.

[8] N. Löw, J. Hesser, M. Blessing, Multiple retrieval case-based reasoning for incomplete datasets, J. Biomed. Inform. 92 (2019), https://doi.org/10.1016/j.jbi.2019.103127.

[9] S.Y. Seo, S.D. Kim, S.C. Jo, Utilizing case-based reasoning for consumer choice prediction based on the similarity of compared alternative sets, J. Asian Finance Econ. Bus. 7 (2) (2020) 221–228, https://doi.org/10.13106/jafeb.2020.vol7.no2.221.

[10] X. Xie, L. Lin, S. Zhong, Handling missing values and unmatched features in a CBR system for hydro-generator design, CAD Comput. Aided Design 45 (6) (2013) 963–976, https://doi.org/10.1016/j.cad.2013.02.004.

[11] J.A. Recio-García, B. Díaz-Agudo, A. Kazemi, J.L. Jorro, A data-driven predictive system using case-based reasoning for the configuration of device-assisted back pain therapy. J. Exp. Theor. Artif. Intell. 33 (4) (2021) 617–635, https://doi.org/10.1080/0952813X.2019.1704441.

[12] O.N. Oyelade, A.E. Ezugwu, A case-based reasoning framework for early detection and diagnosis of novel coronavirus, Inform. Med. Unlock. 20 (2020) 100395, https://doi.org/10.1016/j.imu.2020.100395.

[13] S.V. Shokouhi, P. Skalle, A. Aamodt, An overview of case-based reasoning applications in drilling engineering, Artif. Intell. Rev. 41 (3) (2014) 317–329, https://doi.org/10.1007/s10462-011-9310-2.

[14] H. Zhao, J. Liu, W. Dong, X. Sun, Y. Ji, An improved case-based reasoning method and its application on fault diagnosis of Tennessee Eastman process, Neurocomputing 249 (2017) 266–276, https://doi.org/10.1016/j.neucom.2017.04.022.

[15] A. Patel, O.V. Swathika, U. Subramaniam, T.S. Babu, A. Tripathi, S. Nag, A. Karthick, M. Muhibbullah, A practical approach for predicting power in a small-scale

off-grid photovoltaic system using machine learning algorithms, Int. J. Photoenergy 2022 (2022), Article ID 9194537, 21 pp.

[16] J. Lieber, E. Nauer, H. Prade, G. Richard, Making the best of cases by approximation, interpolation and extrapolation. in: Lecture Notes in Computer Science (including subseries Lecture Notes in Artificial Intelligence and Lecture Notes in Bioinformatics), vol. 11156, (LNAI), Springer Verlag, 2018, pp. 580–596, https://doi.org/10.1007/978-3-030-01081-2_38.

[17] J. Henriet, P.E. Leni, R. Laurent, M. Salomon, Case-based reasoning adaptation of numerical representations of human organs by interpolation, Expert Syst. Appl. 41 (2) (2014) 260–266, https://doi.org/10.1016/j.eswa.2013.05.064.

[18] B. Knight, M. Petridis, F.L. Woon, Case Selection and Interpolation in CBR Retrieval, BCS Spec, Gr. Artif. Intell. 10 (1) (2010) 31–38 [Online]. Available: http://arxiv.org/abs/1509.09199%0Ahttps://doi.org/10.13140/RG.2.1.1387.0800.

[19] P. David, M. Alan, Artificial Intelligence: Foundations of Computational Agents, 2e, 2021. https://artint.info/, (accessed Nov. 02, 2021).

[20] S.H. El-Sappagh, M. Elmogy, Case based reasoning: case representation methodologies, Int. J. Adv. Comput. Sci. Appl. 6 (11) (2015) 192–208. https://doi.org/10.14569/IJACSA.2015.061126.

[21] V. Kurbalija, M. Ivanović, Z. Budimac, Case-based reasoning framework for generating wide-range decision support systems, in: International Conference on Artificial Intelligence and Pattern Recognition 2008, AIPR 2008, 2008, pp. 273–279.

[22] O.V.G.Swathika, K.T.U. Hemapala, IOT based energy management system for standalone PV systems. J. Electr. Eng. Technol. 14 (2019) 1811–1821, https://doi.org/10.1007/s42835-019-00193-y.

[23] J. Howbert, Introduction to Machine Learning & Case-Based Reasoning, in: Introduction to Machine Learning (Course 395), (2012), 20.

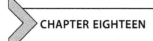

CHAPTER EIGHTEEN

Implementing an ANN model and relative importance for predicting the under drained shear strength of fine-grained soil

Tammineni Gnananandarao[a], Vishwas Nandkishor Khatri[b], Kennedy C. Onyelowe[c,d], and Ahmed M. Ebid[e]

[a]Department of Civil Engineering, Aditya College of Engineering and Technology, Surampalem, Andhra Pradesh, India
[b]Civil Engineering Department, IIT Dhanbad, Jharkand, India
[c]Department of Civil Engineering, Michael Okpara University of Agriculture, Umudike, Nigeria
[d]Department of Civil and Mechanical Engineering, Kampala International University, Kampala, Uganda
[e]Department of Structure Engineering & Construction Management, Faculty of Engineering, Future University in Egypt, New Cairo, Egypt

1. Introduction

The undrained shear strength (S_u) of soil is a widely used design parameter in geotechnical engineering; however, the value obtained depends on the testing apparatus and procedure used [1] as well as the direction of loading, boundary conditions, stress level, sample disturbance, testing method (failure mode), strain rate, stress path, and other factors [2,3]. Uniaxial, triaxial, and direct shear tests on undisturbed samples are routine laboratory tests for the determination of (S_u), whereas the field vane shear test is an in situ test method favored by many. In fact, for a specific soil, the measured shear strength from the in situ test and laboratory tests may give different results for various reasons. To avoid the prediction accuracy, in the present study, ANN models will be developed to increase the accuracy of the prediction. In the past few decades, the ANN applications have come into the light in the geotechnical engineering due to their precision in the prediction of the desired output from the nonlinear data [1,2]. Hence, in the present study, ANN was used to predict the undrained shear strength (S_u) of fine-grained soil using the in situ and laboratory experimental data.

Artificial Intelligence and Machine Learning in Smart City Planning
https://doi.org/10.1016/B978-0-323-99503-0.00012-0

2. Background

Artificial neural networks (ANNs) have been adopted in field of Geotechnical engineering in the early 1990s. In the past few decades, researchers have been effectively applied the ANNs to solve the geotechnical engineering problems (for nonlinear data) such as bearing capacity of H-shaped skirted footing [3] swell pressure vs soil suction behavior [4], shallow footing settlement [5], frictional resistance of drilled shafts for undrained condition [6], bearing capacity of skirted footings [7], bearing capacity and elastic settlement for shallow footings [8], permeability of soils in terms of coefficient [9], strip footing bearing capacity factor [10], slope stability numbers for two-layered soil [11], bearing capacity of eccentrically loaded rectangular foundation resting on reinforced sand [12], power density analysis for microbial fuel cell [13], erodibility of treated unsaturated soil [14], and CPT cone side resistance [15]. In addition, many other (AI) techniques, such as GA, GP, EPR, and PSO, were used to predict the undrained shear strength [16,17] The ANN application is not only limited to the geotechnical engineering, and it has been explored in the other field of civil engineering. From the study of the above literature, it can be concluded that the ANNs can be used to predict any nonlinear data. Hence, in this research, an attempt was made to predict the under drained shear strength of fine-grained soil using ANNs. In this aspect, undrained shear strength of clay soil data was collected from the published literature (in situ and laboratory). By using the above collected ANN model that was developed to predict the undrained shear strength of fine-grained soil (S_u) was first developed. To predict the desired output, cone point resistance (C_p), cone side resistance (C_s), pore water pressure (u), and plasticity index (PI) were used as input parameters. To see the accuracy of the predicted undrained shear strength (S_u), performance measures were used. Finally, sensitivity analysis was performed to see the influence of each input parameter on the output.

3. ANN "artificial neural networks"

ANN has become a well-developed machine learning model that simulates the thinking behavior of "human brain." It consists majorly of three components: network architecture, learning algorithm, and fluctuating function [18]. These computational models are classified into two algorithms: "feed-forward neural network" (FFNN) algorithm and "recurrent

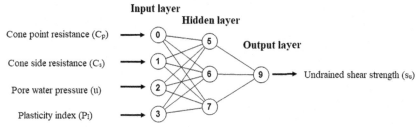

Fig. 1 Architecture of the MLP-FFNN backpropagation algorithms.

artificial neural network" (RANN) algorithm. The FFNN is independent of time; hence, it is suitable when the parameters that used in prediction are not bonded to time [19]. One of the popular FFNNs is the "multilayer perception" (MLP) that consists of many nodes called "neurons" [20,21]. The MLP is usually consists of three layers: input layer, hidden layer, and output layer. Each layer is connected to both previous and next layers with weighted connectors and bias. In the literature [22], it is reported that the MLP is efficient in estimating desired output. Still, prior to predict the desire output ANN requires training. MLP-FFNN is having the different algorithms; however, backpropagation (BP) is considered for the training due to its versatile prediction capability [23]. Also, the performance of the ANN model is dependent on the architecture of the network and the settings of its parameter. The present study architecture of the MLP-FFNN backpropagation algorithms was fixed [2] and shown in Fig. 1. Therefore, neural network model chosen for the present study is 4-3-1 for the constitutive ANN modeling (from Fig. 1).

The main issue in successfully applying the neural network model is to select the right condition to terminate training. Overtraining for ANN generates noise, while undertraining produces poor predictions. As a result, generalization of the network will be absent for a new dataset. As a result, both training and testing dataset iteration numbers were updated using trial and error. For each iteration, MSE between both predicted and actual values was calculated and made a plot for number of epochs and MSE as shown in Fig. 2. From Fig. 2, the smallest SME corresponding epoch 2000 was chosen for the ANN model preparation.

4. Data collection

This article aims form implementing an artificial neural network model and relative importance for predicting the under drained shear

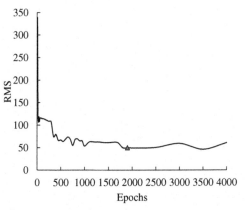

Fig. 2 Assessment of optimum epochs.

Table 1 Data ranges used for artificial neural network model variables.

Parameters (input/output)	Minimum	Maximum	Mean	Standard deviation
Cone point resistance (C_p)	591	2360	1489.16	436.08
Cone side resistance (C_s)	0	121	42	27.9
Pore water pressure (u)	108	488	273.42	104.47
Plasticity index (PI)	8	28	15.84	4.49
Undrained shear strength (S_u)	17.85	104.38	57.08	22.92

strength of fine-grained soil. The test result database considered in this research has 50 records, and they were extracted from the work done by [24]. Table 1 presents this database. To apply ANN modeling, the dataset was randomly splitted to training set and testing set that is used to validate the generalization of the trained ANN model. The training set had 36 records (72% of total data) while the testing set had 14 records (28% of total data).

5. Performance indicators and developed ANNs model

After the data collection, the training was initiated with help of the above fixed (architecture, iteration, and hidden layers) procedure. To ensure the prediction accuracy, performance measures were used. The performance measures were chosen based on the literature [2], for example, "coefficient of determination" (R^2), "correlation coefficient" (r), "mean squared error" (MSE), "root-mean-square error" (RMSE), "mean absolute error" (MAE), and "mean absolute percentage error" (MAPE). The predicted model has r and R^2 values close to 1 and MSE, RMSE, MAE, and MAPE values closer

Table 2 Calculated performance measures values of ANN model.

Performance measures	Training	Testing
r	0.99	0.99
R^2	0.98	0.98
MSE	107.58	101.75
RMSE	10.37	10.09
MAE	8.65	8.63
MAPE (%)	19.58	20.88

to zero, indicating a perfect model [2,25,26]. The performance indicators for both training and testing sets of the proposed mode are shown in Table 2. The study of the Table 2 reveals that the proposed model can predict the undrained shear strength of fine-grained soil in precise manner.

The weight matrix of the proposed ANN model, which contains the connector weights of both input-hidden and hidden-output including the bias, is presented in Eqs. (1)–(4)

$$x_{ji} = \begin{bmatrix} x_{11} & x_{12} & x_{13} & x_{14} \\ x_{21} & x_{22} & x_{23} & x_{24} \\ x_{31} & x_{32} & x_{33} & x_{34} \end{bmatrix} \tag{1}$$

$$= \begin{bmatrix} -0.21 & -0.04 & -0.09 & -0.13 \\ -2.58 & 0.12 & -1.44 & -0.53 \\ 1.28 & 0.05 & 0.92 & 0.33 \end{bmatrix}$$

$$y_{jk} = \begin{bmatrix} y_{11} \\ y_{21} \\ y_{31} \end{bmatrix} = \begin{bmatrix} -0.10 \\ -3.94 \\ 1.94 \end{bmatrix} \tag{2}$$

$$z_j = \begin{bmatrix} z_1 \\ z_2 \\ z_3 \end{bmatrix} = \begin{bmatrix} -0.35 \\ 1.64 \\ 0.67 \end{bmatrix} \tag{3}$$

$$z_0 = [0.52] \tag{4}$$

6. Equivalent equation of ANN model

After developing the final trained weights, a model equation was proposed in this section based as per [26]. Taking into account the weights and biases given by Eqs. (1)–(4), the ANN model takes the following form:

$$s_u = f_n \left\{ z_0 + \sum_{i=1}^{n} \left[y_{jk} f_n \left(\sum_{i=1}^{n} x_{ji} E_i \right) \right] \right\} \tag{5}$$

where,

  h: no. neurons of the hidden layer (=8 for this model)

  E_i: normalized input values (0.0–1.0)

  f_n: the utilized activation function

  n: no. of neurons of input layer

The following steps [$A_1 - A_4$ and $B_1 - B_4$ using Eqs. (6)–(9) and (10)–(13), respectively] were carried out for the development of the model equation using ANN. Eq. (15) provides the output S_u in denormalized form.

$$A_1 = x_{11} \times c_p + x_{12} \times c_s + x_{13} \times u + x_{14} \times PI + z_1 \tag{6}$$

$$A_2 = x_{21} \times c_p + x_{22} \times c_s + x_{23} \times u + x_{24} \times PI + z_2 \tag{7}$$

$$A_3 = x_{31} \times c_p + x_{32} \times c_s + x_{33} \times u + x_{34} \times PI + z_3 \tag{8}$$

$$A_4 = x_{41} \times c_p + x_{42} \times c_s + x_{43} \times u + x_{44} \times PI + z_4 \tag{9}$$

$$B_1 = \frac{y_{11}}{1 + e^{-A_1}} \tag{10}$$

$$B_2 = \frac{y_{21}}{1 + e^{-A_2}} \tag{11}$$

$$B_3 = \frac{y_{31}}{1 + e^{-A_3}} \tag{12}$$

$$B_4 = \frac{y_{41}}{1 + e^{-A_4}} \tag{13}$$

$$s_u \, (\text{kPa}) = B_1 + B_2 + B_3 + B_4 + z_0 \tag{14}$$

After denormalization

$$s_u \, (\text{kPa}) = 0.5(s_u + 1)\left([s_u]_{max} - [s_u]_{min} \right) + [s_u]_{min} \tag{15}$$

From the collected database from previous literature, a model equation using ANN is proposed (Eq. 15). This equation can be used to determine the undrained shear strength of fine-grained soil.

7. Results and discussions

  After developing the model equation for the calculating the undrained shear strength of the fine-grained soil, the predicted vs targeted undrained shear strength graph was drawn for training and testing as depicted in

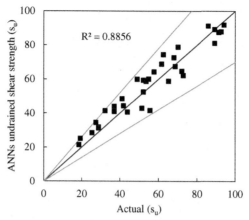

Fig. 3 Variation between actual undrained shear strength (S_u) vs predicted ANN undrained shear strength (S_u) for the training data.

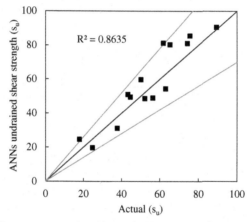

Fig. 4 Variation between actual undrained shear strength (S_u) vs predicted ANN undrained shear strength (S_u) for the testing data.

Figs. 3 and 4, respectively. The study of Figs. 3 and 4 reveals that the R^2 values of the training 0.98 and testing 0.98 are closer to 1. However, the data points are presented with the ±30% error line in the both the training and testing as evident from Figs. 3 and 4. Hence, the proposed equation is having the capability to predict the for the undrained shear strength of fine-grained soil in precise manner.

Finally, the variation of the ANN and actual data were compared for the testing data as presented in Fig. 5. Fig. 5 reveals that the predicted (from ANNs) S_u and actual S_u are not having much variation.

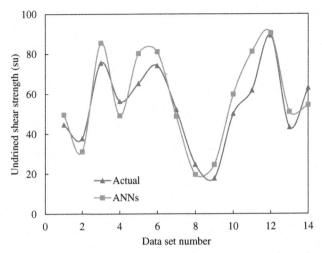

Fig. 5 The plot of agreement between actual and predicted undrained shear strength values using ANNs.

8. Relative importance

This section in this article discusses the involvement of each individual input parameters on the output (undrained shear strength of the fine-grained soil) by conducting the relative importance study. For this subject, the technique that depends on weight configuration reported by [27] was used. The results obtained from the relative importance method are illustrated in Fig. 6, which indicates that, the cone point resistance (C_p) is most influenced input parameter in predicting the output followed by plasticity index (PI), cone side resistance (C_s) and pore water pressure (u) having the 45.46%, 26.62%, 19.76%, and 8.16%, respectively.

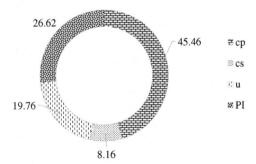

Fig. 6 Plot of relative importance.

9. Conclusions

This article is aimed at developing the model equation for the undrained shear strength of the fine-grained soil using ANNs with backpropagation algorithm. To achieve the model equations, the independent variables used were cone point resistance (C_p), cone side resistance (C_s), pore water pressure (u), and plasticity index (PI). The proposed expression of the undrained shear strength of the fine-grained soil was made using ANNs with a backpropagation algorithm, and the following conclusions were drawn:

1. The developed ANN model can predict the undrained shear strength of the fine-grained soil in precise manner (error between the actual vs predicted is within the ±30).

2. The relative importance is conducted using the weight reveals that the cone point resistance (C_p) is most influenced input parameter in predicting the output followed by plasticity index (PI), cone side resistance (C_s), and pore water pressure (u) having the 45.46%, 26.62%, 19.76%, and 8.16%, respectively.

3. The calculated performance indicators (r, R^2, MSE, RMSE, MAE, and MAPE) result in indicating that the developed ANN model is quite good in predicting the output.

Overall, the paper has made an effort to provide the understanding of ANNs to predict the undrained shear strength of the fine-grained soil. The ANN-based equation will be useful for calculating the undrained shear strength of the fine-grained soil which otherwise need expensive investigation.

References

[1] R.K. Dutta, R. Rani, T. Gnananandarao, Prediction of ultimate bearing capacity of skirted footing resting on sand using artificial neural networks, J. Soft Comput. Civil Eng. 2 (4) (2018) 34–46, https://doi.org/10.22115/SCCE.2018.133742.1066.

[2] R.K. Dutta, K. Dutta, S. Jeevanandham, Prediction of deviator stress of sand reinforced with waste plastic strips using neural network, Int. J. Geosynth. Ground Eng. 1 (2) (2015) 1–12.

[3] T. Gnananandarao, V.N. Khatri, R.K. Dutta, Prediction of bearing capacity of H shaped skirted footings on sand using soft computing techniques, Arch. Mater. Sci. Eng. 103 (2) (2020) 62–74.

[4] Y. Erzin, Artificial neural networks approach for swell pressure versus soil suction behavior, Can. Geotech. J. 44 (10) (2007) 1215–1223, https://doi.org/10.1139/T07-052.

[5] T. Gnananandarao, R.K. Dutta, V.N. Khatri, Application of artificial neural network to predict the settlement of shallow foundations on cohesionless soils, Geotechn. Appl. Lecture Notes Civil Eng. 1 (3) (2019) 51–58, https://doi.org/10.1007/978-981-13-0368-5_6.

[6] A.T.C. Goh, F.H. Kulhawy, C.G. Chua, Bayesian neural network analysis of undrained side resistance of drilled shafts, J. Geotech. Geoenviron. 131 (1) (2005) 84–93, https://doi.org/10.1061/(ASCE)1090-0241(2005)131:1(84).

[7] E. Momeni, D.J. Armaghani, S.A. Fatemi, R. Nazir, Prediction of bearing capacity of thin-walled foundation: a simulation approach, Eng. Comput. 34 (2) (2018) 319–327, https://doi.org/10.1007/s00366-017-0542-x.

[8] M. Omar, K. Hamad, M.A.I. Suwaidi, A. Shanableh, Developing artificial neural network models to predict allowable bearing capacity and elastic settlement of shallow foundation in Sharjah, United Arab Emirates, Arab. J. Geosci. 11/16 (2018) 464, https://doi.org/10.1007/s12517-018-3828-4.

[9] H.I. Park, Development of neural network model to estimate the permeability coefficient of soils, Mar. Georesour. Geotechnol. 29 (4) (2011) 267–278, https://doi.org/10.1080/1064119X.2011.554963.

[10] R. Sahu, C.R. Patra, N. Sivakugan, B.M. Das, Use of ANN and neuro fuzzy model to predict bearing capacity factor of strip footing resting on reinforced sand and subjected to inclined loading, Int. J. Geosynth. Ground Eng. 3 (2017) 1–15, https://doi.org/10.1007/s40891-017-0102-x.

[11] P. Samui, B. Kumar, Artificial neural network prediction of stability numbers for two-layered slopes with associated flow rule, Electron. J. Geotech. Eng. (2006). http://eprints.iisc.ac.in/id/eprint/5750.

[12] R. Sahu, C.R. Patra, K. Sobhan, B.M. Das, Ultimate bearing capacity prediction of eccentrically loaded rectangular foundation on reinforced sand by ANN, in: International Congress and Exhibition. Sustainable Civil Infrastructures: Innovative Infrastructure Geotechnology, 2018, pp. 45–58.

[13] K. Singh, Dharmendra, Power density analysis by using soft computing techniques for microbial fuel cell, J. Environ. Treat. Techn. (2019) 1068–1073.

[14] K.C. Onyelowe, T. Gnananandarao, C. Nwa-David, Sensitivity analysis and prediction of erodibility of treated unsaturated soil modified with nanostructured fines of quarry dust using novel artificial neural network, Nanotechnol. Environ. Eng. 6 (2021) 37, https://doi.org/10.1007/s41204-021-00131-2.

[15] T. Gnananandarao, R.K. Dutta, V.N. Kahtri, Neural network based prediction of cone side resistance for cohesive soils, Lecture Note Civil Eng. 137 (2021) 389–399, https://doi.org/10.1007/978-981-33-6466-0_36.

[16] A.M. Ebid, 35 years of (AI) in geotechnical engineering: state of the art, Geotech. Geol. Eng. (2020), https://doi.org/10.1007/s10706-020-01536-7.

[17] A.H. El-Bosraty, A.M. Ebid, A.L. Fayed, Estimation of the undrained shear strength of east Port-Said clay using the genetic programming, Ain Shams Eng. J. (2020), https://doi.org/10.1016/j.asej.2020.02.007.

[18] P.K. Simpson, Artificial Neural System: Foundation, Para- Digms, Applications and Implementations, Pergamon, New York, 1990.

[19] M.A. Shahin, H.R. Maier, M.B. Jaksa, Predicting settlement of shallow foundations using neural networks, J. Geotech. Geoenviron. Eng. 128 (9) (2002) 785–793.

[20] S. Haykin, Neural Networks, second ed., Prentice-Hall, Englewood Cliffs, 1999.

[21] M. Rezaei, M. Monjezi, S.G. Moghaddam, F. Farzaneh, Burden prediction in blasting operation using rock geomechanical properties, Arab. J. Geosci. 5 (2012) 1031–1037.

[22] K.L. Du, A.K.Y. Lai, K.K.M. Cheng, M.N.S. Swamy, Neural methods for antenna array signal processing: a review, Signal Process. 82 (2002) 547–561.

[23] G. Dreyfus, Neural Networks: Methodology and Application, Springer, Berlin, 2005.

[24] A. Cheshomi, Empirical relationships of CPTu results and undrained shear strength, J. Geo Eng. 13 (2) (2018) 49–57, https://doi.org/10.6310/jog.201806_13(2).1.

[25] T. Gnananandarao, R.K. Dutta, V.N. Khatri, Bearing capacity and settlement prediction of multi-edge skirted footings resting on sand, Ing. Investig. J. 40 (3) (2020) 9–21.

[26] T. Gnananandarao, V.N. Khatri, R.K. Dutta, Prediction of bearing capacity of H plan shaped skirted footing on sand using soft computing techniques, Arch. Mater. Sci. Eng. 2 (103) (2020) 62–74, https://doi.org/10.5604/01.3001.0014.3356.

[27] G.D. Garson, Interpreting neural-network connection weights, AI Expert. 6 (4) (1991) 46–51.

PART FIVE

Smart transportation

CHAPTER NINETEEN

Smart transportation based on AI and ML technology

Swetha Shekarappa G.[a], Sheila Mahapatra[a], Saurav Raj[b], and Manjulata Badi[a]

[a]Department of Electrical and Electronics Engineering, Alliance University, Bangalore, Karnataka, India
[b]Department of Electrical Engineering, Institute of Chemical Technology, Jalna, India

1. Introduction

People on public transportation face several problems with collecting fares, especially from the customers while inspecting a travel pass becomes tricky when the congestion that is the number of passengers is very large in numbers. Whether a city's transport system is efficient determines the productivity of the city. In order to get to their jobs, they are required to travel every day and face this problem. In this chapter, a clever method is designed to deal with the collection problems associated with the fare mentioned earlier. A valuable system such as this enhances the fare collection and congestion management process. Making efficient and effective decisions is possible by employing artificial intelligence (AI) techniques such as machine learning (ML). Given the presence of inexpensive and capable machines, the use of machine learning (ML) has spread widely across many different areas, including robotics, business, the arts, and automated systems. ML assists the creation of smart and efficient decision-making for overall system performance, which can include the ability to be reliable, conserve energy and ensure that a quality level of service is provided (QoS).

With population growth and the proliferation of vehicles, traffic congestion and safety have become increasingly challenging in many urban areas. Road accidents cause the deaths of more than one million people each year. Congestion leads to unnecessary costs, such as delays, stress, and pollution, as well as waste of fuel. The congestion cost in the United States in 2017 was $305 billion, which was derived from literature. A successful transport system is able to provide increased traffic flow while reducing traffic accidents and resulting in a cleaner environment. The VANET is designed to improve traffic and safety inefficiencies in order to reduce travel time, especially in peak times. The proliferation of connected devices that make up the Internet of things (IoT) has caused applications to become more intelligent and allow for their exploitation in various aspects of modern cities. The power of machine learning (ML) increases as more data are added to the machine learning database. Much work has been done in the field of smart transportation, and this research has attracted both ML and IoT practitioners in the literature. Conducting more research into intelligent transport system (ITS) applications such as route optimization, parking, and accident/detection tends to be among the most popular ITS research endeavors.

This paper's goal is to convey several intelligent technologies' most important characteristics, along with the services they are compatible with for development of new capabilities. In order to highlight and prove the proposed ideas and concepts, an AI-based transportation industry architecture is put forward.

Seventy percent of the world's population is expected to live in cities by the year 2050, according to [1]. As a result, cities are essential to ensuring the long-term sustainability of the world. As a result of the environmental challenges and opportunities, this movement poses for everyone including individuals, businesses, nonprofits, and governments. There are still questions about the ultimate goal of improving living conditions for everyone. The United Nations created the 2030 Agenda for Sustainable Development in order to achieve SDG 11, Sustainable Cities and Communities, Creating safe, resilient, and long-lasting cities and human settlements. There will be a need for new knowledge and technological advancement just like in the past. Step back and assess the current state of affairs. In order to promote inclusive, tolerant, and open cities, the focus on "smart cities" must include both nature and the physical environment as well as citizen education. As the 2030 Agenda for the United Nations, in particular Sustainable Development Goal SGD 11, has evolved, the concept of a "smart city" has emerged. The UN's vision puts people first at every stage of the built environment's lifespan, from inclusion and safety to preparing for unattended but simulated extreme events. Cities, as a whole, must be sustainable and provide citizens with a high quality of life today and in the future, while also taking into consideration the natural environment. When it comes to creating "Smart and Liveable Cities," the most important question a city must ask itself is whether or not it can get its citizens involved in the process of integrating digital and physical infrastructures seamlessly with citizens' involvement. One of the primary objectives of the smart city development is to provide citizens with an environmentally sound and economically efficient built environment that includes transportation infrastructures like roads and mass transit networks, all while increasing productivity. Increased productivity and competitiveness; a reduction in the city's environmental impact, and a better quality of life for its residents are all the result of this strategy. For this study, the most important aspects of smart cities and how transportation systems can maximize the benefits that technology implementation brings to towns and cities [2] are examined, which could be replicated around the world in a variety of ways. According to this research, smart cities are successful in achieving and promoting the three sustainability pillars (environmental, economic, and social). When it comes to transportation, this article begins with a review of the literature on smart city components, which is followed by a discussion of this topic that emphasizes the importance of technology in transportation systems and the application of MaaS to encourage environmentally friendly and economically efficient travel within the city.

2. Review of related literature

Various buzzwords for better-than-average cities have emerged, such as "smart," "sustainable," "resilient," and "liveable." However, the exact meanings of these buzzwords vary depending on who uses them. While cities and regions are "living organisms," they should be viewed as a constant source of growth and renewal [3,4]. They should be seen as ready to welcome new activities into the area at any given time. "Smart cities" have been the subject of much discussion in recent years, and it is widely assumed that they are technology-based systems. This system is supported by major data analytics and artificial intelligence, with multiple interconnected layers corresponding to various components of the city environment [4]. It's depicted in Fig. 1. This program's ultimate goal ought to be to make it easier for city officials and residents to fulfill their civic duties. It's important to ask what a truly great city would look like today and in the future for everyone when considering future city concepts. [5]. To attract new businesses and institutions, provide well-paying jobs to current employees, and attract new talent, the city should be guided by sound and trustworthy governance [6]. When it comes to competing in today and tomorrow's economy, cities will be judged on how well they are able to maintain a sustainable and resilient built environment, preserve their historic structures, operate city systems and services, as well as respond quickly to challenges and threats. Maintaining a harmonious relationship between the city and its natural

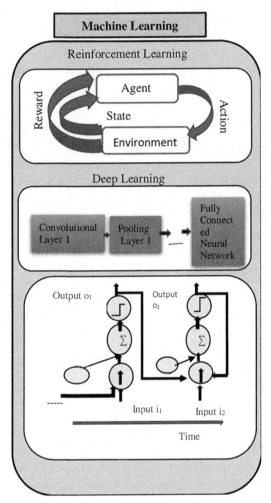

Fig. 1 AI techniques (ML) used in EV and mass adoption.

surroundings will be a government responsibility for the foreseeable future, so it is imperative that the competitive city is both pleasant and healthy, as well as mindful of its natural surroundings. To achieve a competitive and efficient city, a digital transformation will be an essential part of the project, without ignoring the security of systems and people [7].

Creating a competitive and efficient city requires a digital transformation project that does not overlook the importance of cybersecurity for both systems and people. In order to achieve a competitive city that is both livable and sustainable, governments must play an integral role in governance, with citizens, businesses, research institutions, and technological institutions all playing a role. It will be possible for every generation in the city to participate in the creation of their city's visions.

3. The built environment's role in creating an efficient and competitive city

If the best materials and technologies are used in conjunction with intelligent transportation systems (ITS), as well as user input and consideration of the life cycle assessment (LCA), which takes into account embodied energy not only during the use phases of the built environment, but also during all processes from extraction to redevelopment, the built environment can be considered sound. The construction industry will become a service-oriented business in the future if it adopts flexible and connected spaces and inclusive cities with a high quality of life for all citizens are provided by a user-centric approach. Buildings powered by renewable energy should prioritize the needs of their occupants in terms of their well-being, happiness, and efficiency. They should be able to incorporate renewable energy sources, storage, data, and digital technologies [8]. In order to provide a seamless experience for the occupants, they should also be connected to the surrounding transportation system, particularly the highway system. This diagram illustrates how the government should work with citizens, businesses, and research and development institutions to improve governance. The government has the unquestionable responsibility of fostering leadership that aids cities in competing and thriving [9]. Throughout the generations, citizens, from the youngest to the oldest, will be able to participate in the construction of their city's visions.

With the implementation of a smart grid, cities will be able to better manage their energy needs, protect the distribution network, and save money and resources. In order to integrate intelligence into the physical environment, city-wide digital transformations will allow electronic and digital systems to be embedded in buildings and infrastructure [10]. The Internet of Things, or IoT, refers to the billions of physical devices around the world that are now connected to the Internet, so that the communication wirelessly. In order to provide a rich context for analysis, it is critical to store and process a large amount of heterogeneous data flowing at various rates. Big data can be difficult to manage, especially if you're still relying on out-of-date methods and tools. What is the role of the citizens in this complicated process? Finally, their hard work will pay off and raise their standard of living. Citizen participation as data providers and end users is only possible if citizens feel prepared and involved in all processes through adequate digital literacy [11]. To spread the word about smart technology, people need to know what they want and have faith in Internet of things devices. [12]. City governance is a hugely complex process that involves multiple agencies and stakeholder groups (governments, businesses, citizens) working together to resolve conflicts of interest [13]. In order to maximize the socioeconomic and environmental performance of these smart cities, knowledge transfer facilitates decision-making by connecting all the forces at work [14].

4. Current challenges in mobility and transportation systems

There has been an increase in transportation demand, which means that private cars have become the dominant mode of urban mobility in large cities [15]. Traffic congestion, environmental pollution (such as noise), and traffic accidents are all exacerbated by an increase in road use. Urban mobility services are negatively affected by these factors, which have a direct impact on accessibility and urban mobility [16]. Delays, energy consumption, pollution, and stress are all exacerbated as a result, which lowers productivity and raises overall societal costs. There is still a lot of concern about car accidents. More than 20,000 people die and more than 1lakh people are seriously injured every year in the EU-28 [17]. Mobility serves as the primary driver of sustainability and the primary integrator of objects and people, and as a support for the flow of information in urban space as a proxy for human activity. In order to ensure timely and environmentally friendly mobility, the transportation system must incorporate new modes of transportation and take into account evolving user demands, especially for those who are young, elderly, and/or less able. These are the people on the road who are most at risk (VRUs). Transport systems must also adapt to mixed traffic environments where current and future connected and automated vehicles co-exist (CAVs). The European Commission sees the introduction of CAVs as critical to GHG emissions will be reduced by 60% and traffic fatalities will be nearly eliminated by 2050. Autonomous vehicles must first overcome the difficulties of living alongside conventional transportation methods in mixed traffic environments where people have a "respect for, acceptance, and reliance on automation" as well as "the role of humans when necessary." Any challenges that may arise necessitate the availability of up-to-date and relevant data by using innovative digital technologies in a multidisciplinary environment that includes mobile technology, sensorization in transportation, and mobility infrastructure components, not only increases vehicle data collection capacity and intelligence but also takes into account the needs of users by employing a global system of sensors in these components. System data collection is required to support the production of essential information for the integrated management of mobility and to provide answers to global decisions about the city environment for this system (noise, air quality). Other uses include developing key performance indicators for various areas of decision-making, such as the current mobility operation and optimization of the global transportation system and developing alerts for serious threats. Mobility disruption can be caused by natural and man-made disasters, including hurricanes, flooding, and fires, as well as cultural events, such as protests and labor disputes. Human error or mismanagement can also lead to system failures. Resilience can only be fully determined if transportation systems are tested for their ability to absorb and/or reduce the effects of various disturbances and maintain or restore normal operation within a reasonable time and cost framework. Digitalization in transportation has the potential to provide passengers with highly personalized services and commercial offerings in the long term. Users' impact and their readiness to take advantage of new opportunities are the primary sources of mobility disruption. On the other hand, they are crucial and frequently overlooked despite this. A "building capacity" approach to data governance teaches users how to trust the global process of capturing, treating, and producing information so that they can freely share their data with the rest of the community. To ensure that all users have an equal opportunity to understand and contribute to the development of new mobility management features, different groups of users should be considered. Urban living labs would be a good place to test and innovate new technology, devices, and systems because they are both robust and efficient. AI has the potential to make electric vehicles more safe, reliable, and economically viable, all of which are necessary for their widespread adoption. As an example, the design and manufacture of electric vehicle battery packs has been extensively studied, as has the optimization of EV range and the design of EV controls. Rechargeable batteries management and range enhancement are two areas in which AI can be used to improve energy efficiency.

5. ML is being used in battery research and development

Electric vehicles rely heavily on lithium batteries for as well as a means of storing energy [18]. To create a larger battery pack, electric vehicle batteries are interconnected and assembled into modules. It's possible to connect these batteries in a series or parallel configuration to get the power and capacity you need. Due to design and manufacturing limitations, current EV batteries still have performance-related issues that affect their performance while driving. Design and manufacturing restrictions on electric vehicle (EV) batteries have led to their lower energy density [19]. In addition, more efficient electric vehicle batteries can greatly reduce the fear of running out of juice [20]. Additionally, the ability of ML to speed up the process of finding and describing new things as well as improvement of battery materials and manufacturing. A higher level of battery efficiency and safety can be achieved through increased efficiency.

Quantum mechanics (QM), computational advances, and materials datasets have made battery material discovery and characterization easier and more affordable with machine learning (ML) [21,22]. Battery electrodes, solid electrolytes, and electrolyte additives are examples of materials whose energy density and safety can be improved using machine learning techniques (ML) [23,24].

According to a recent study, machine learning (ML) can be used to predict the anticipated characteristics of new materials from the database's existing materials [25]. These unknown or difficult-to-measure properties of existing materials can be estimated using these data that are then fed into machine learning techniques (ML) [26,27] via QM calculations or databases (such as QM9) and tomography can also speed up the discovery of new materials. Dataset size [28] and industry reluctance to switch to new battery materials have stifled commercial interest in ML-based battery design approaches [29], which have been demonstrated in numerous studies [30]. Small datasets are often used by researchers to study difficult-to-quantify properties of materials [31]. Due to their limited size, small datasets have a negative impact on ML's ability to select and utilize accurate models. Developing new materials quickly isn't the only problem; there aren't enough people with the necessary training and experience to put them to use in industry [32]. When battery electrodes, for example, are manufactured with their properties altered to increase the battery's energy density even further in [33]. As a result, the battery's theoretical maximum energy density is maintained as close to its actual value as possible. There are many ways to improve batteries, including increasing the surface area of their electrodes [34]. Predicting intermediate product characteristics from manufacturing process parameters has been demonstrated by recent studies. As an example of a mixer, if we add some chemical composition to its electrode then the speed can be enhanced. To illustrate this point, mass loading of electrodes is predicted using ML regression-based models for control parameters, such as mass content, solid–liquid ratio, and viscosity in electrode fabrication [35]. There aren't many relevant datasets to draw from because it's a new field of study. For research purposes, creating datasets necessitates a substantial investment of time and resources. Due to the lack of input features, the dataset generated by is of little use to researchers. As this field's research and commercial activities expand, more data may become available. Empirical and physical models can be used to use to create simulations that quickly generate datasets. A computer-generated neural network charge-discharge specific resistance can be predicted using simulations rely solely on the accuracy of simulations. This approach is labor- and cost-intensive, but it yields a substantial dataset. Combining experimental results with model-based simulators ensures high reliability and large datasets for machine learning algorithm training [36]. In addition, more research is needed to examine the ML-based insights into the relationship between the production process and the final product's output.

6. Artificial intelligence (AI) role in electric vehicle (EV) and smart grid integration

6.1 An overview of the integration process and the primary issues it faces

This two-way flow of energy is not possible with traditional electric power distribution systems. Renewable energy production and storage integration, such as charging an EV at a price based on supply and demand, can't be handled by them [37]. G2V refers to a flow of energy from the grid to an electric vehicle (EV), while V2G refers to a flow of energy back to the grid. Thus, from the point of view of a smart grid, an electric vehicle can be seen as both an energy store and a power generator (V2G) [38]. An efficient grid–energy generation and distribution by the fact that this bidirectional flow of energy allows for frequency regulation, peak shaving and loading leveling, and

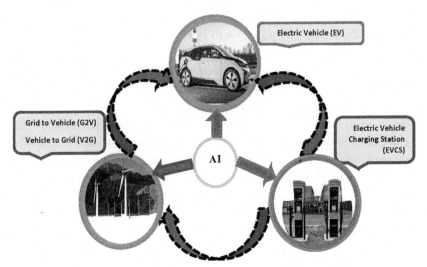

Fig. 2 Overview of AI in EV, EVCS, and EV integration with smart grid.

spinning reserve (Fig. 2), particularly when intermittent sustainable energy sources like solar and wind are used.

There is a lot of interest in V2G technology because of its ability to generate energy for the grid [39]. There are many advantages to V2G but there are also many disadvantages, including technical and societal resistance [40], high initial investment costs and difficulties in distributing energy. Eco-friendly smart grid design increases profits and reduces operating expenses in an electric vehicle [41]. Artificial intelligence (AI) can be a helpful tool in dealing with these issues. An AI-powered system is necessary to ensure smooth power distribution because of the intermittent nature of renewable energy.

6.2 The production and distribution of electricity should be optimized

It's important to keep in mind that there are a number of factors that limit the amount of power that can be generated and distributed. It has been found that combining computer intelligence with ML is an effective way to increase the efficiency of power generation and distribution increase revenue for electric vehicle users while minimizing overall operating costs. Regulations for demand and supply, generation, voltage and thermal limits and EV charging limit were put in place. PSO and the GA simulation were used to solve MOOP in a 33-bus distribution network and 1800 EVs. As an added bonus, researchers found a way to minimize operational costs while also improving the lifespan of electric vehicle batteries and reducing wait times in the EVCS queue. As a result of a battery cost model that incorporates state-of-charge data (AFSA), artificial fish swarms were used to solve the MOOP [42]. By adjusting the model parameters, battery degradation costs were slashed by 80% and queue wait times by 60% [43,44]. It was because of this that machine learning to analyze the factors that influence users' decision-making and their ability to follow through. The availability and location of EV users were tracked using an app, which was used as a model input. Predictions of an EV fleet's presence at charging stations for four weeks were accurate to within 5% of the time [45]. In order to determine how much storage space is available, this model is essential. Optimization of renewable energy systems, it is possible to apply the same energy management and optimization techniques to wind turbines and solar panels are examples of renewable energy generation systems. Electrochemical storage capacity of electric vehicles can be used in particular for intermittent renewable power generation. Examples of how V2G and renewable energy sources that can be integrated into the power generation system are provided below to keep operating costs low and maximize profits for electric vehicle (EV) owners [46]. PSO outperformed mixed integer programming linear programming by 148 times in a 32-bus distribution network with 50 plug-in hybrids. PSO (power system optimization) is used to solve a MOOP while reducing CO2 emissions and wind curtailment [47]. For the Pareto front, decision-based fuzzy rules were employed. Reduce CO_2 emissions by synchronizing the discharging of electric vehicles at high demand and low wind speed times. Electric vehicles can reduce wind curtailments by charging at times when dealing with wind imbalances and V2G costs, PSO and Pareto front are used to solve for MOOP costs. For balancing out imbalances in wind power generated by a 10-MW wind plant and a 484-vehicle electric vehicle fleet, V2G outperforms gas generators and battery systems.

7. Smart warehouse logistics and supply chain management

Our final look at the landscape includes smart logistics and the smart warehouse, both of which have recently emerged the relevant papers.

7.1 Utilization of intelligent logistics

Growing fleets, complicated transportation networks, and a widening range of delivery requirements [48] would all be addressed by smart logistics, according to their predictions [49]. The objective was to meet delivery needs while also being environmentally friendly through a variety of information and communication strategies. A framework for making decisions about on-demand parcel delivery while also considering just-in-time, fuel consumption, and carbon emissions. They came up with an algorithm that combined static and dynamic elements as a solution. There were fuel consumption reductions of 6.4% and a 2.5% reduction in carbon dioxide emissions thanks to the framework. IoT technology was used to gather real-time data, which was then shared among logistics companies to enable a dynamic optimization strategy driven by real-time information for smart vehicles and logistics tasks after that study was completed [49]. Reduced fuel consumption and lower logistics costs were the ultimate objectives.

The dynamic optimization method was used as part of the optimization process in order to reduce carbon emissions. Smart logistics has been implemented in China, which is really impressive [50]. They examined the effects of smart logistic policy on freight volume, logistics employment, and total social retail sales using a model based on the difference-in-differences principle. There are three main components to the smart logistics ecological chain (SLEC), which is a new concept developed from smart logistics: a supply ecological group, a core ecological logistics platform, and an eco-demand ecological group [51]. Logistics play a critical role in bridging the gap between upstream and downstream groups. As a result, it was determined that SLEC implementation requires both empowerment capacity and information sharing [52].

7.2 A storage facility equipped with cutting-edge technology

A number of papers use the term "smart warehouse" in the same way that "smart logistics" describes warehouses that are enabled by automation and various technologies in order to improve operational efficiency and decrease latency. Customers' orders are treated differently depending on the level of service they require. As a solution to this issue, the differentiated probabilistic queueing policy was put forth as an option. The alternating minimization method they used to fix the problem resulted in a performance boost of 19.64%. Artificial intelligence in warehouses is a hot topic at the moment, and Buntak et al. [53] examined the potential pitfalls and rewards of implementing it. The near-term potential of AI may be overestimated by implementation and tech groups, while management with limited exposure to AI will overlook the issue. Like smart logistics, the environment is a major concern as well. In smart warehouses, industrial cooling systems use an artificial neural network to learn the best operational parameters. When comparing the new approach with the old, they discovered a 3.2% improvement in performance.

8. Conclusions and future directions are summarized in this section

There is no doubt that stem cells have been a hot topic and have grown in popularity in recent years. Additionally, applications for manufacturing floors, warehouses, and last mile deliveries have been developed. One of the many advantages and contributions that autonomy offers is the ability to operate more efficiently and at higher levels of service thanks to AI/BD/ML with IoT/blockchain support and other emerging technologies. Numerous advantages have been demonstrated, but optimizations face numerous difficulties, such as the conversion of data into meaningful optimization parameters and so on. It is anticipated that STs will be essential in a wide range of future applications.

8.1 Blockchain-based digitalization of supply chain logistics

Many shipping and logistics companies still rely heavily on paper for various verifications and records, resulting in low operational efficiency and high operating costs. To replace and improve operational efficiency and information sharing, the concept of logistics digitalization is being promoted and rapidly developed. Verifying the bill of lading is one such example. Freight rates, shipper requirements, and distribution center storage capacity are just a few examples of the types of sensitive and confidential information that goes into logistical business operations. As a result, it is becoming increasingly difficult to maintain data integrity and accuracy. According to recent research, blockchain technology can be a useful tool in the fight against dishonesty [54]. As a result, moral hazard concerns are minimized and fair trade is preserved.

8.2 Reducing the cost of personal transportation by pooling resources

The use of autonomous vehicles (AVs) to replace existing public transportation, such as the public bus/shuttle system, shows great potential, as evidenced by the papers we've examined. Adding AVs to public transportation systems during rush hours has been shown in some studies to reduce passenger wait times and thus improve service quality. At this point, research is focused on a single route rather than the actual operation mechanism itself, which is still in the conceptual stage. Passing the optimization's complexity is another challenge. Prior to meeting the demands of a real-world transportation network with multiple routes, more research in this area is required.

8.3 Internal logistical integration that is both intelligent and seamless

Currently, most of the work in intelligent manufacturing and storage is done in silos. Using AR to automate shop floors, smart manufacturing aims to become more efficient, while the smart warehouse uses robotics to automate the storage and retrieval of items. Materials are stored in warehouses before being shipped to the production floor for assembly. After that, the completed or partially completed item will be returned to the storage facility for safekeeping, where it will remain for a period of time until needed again. In terms of logistics, there is a lot of back and forth between the two locations. To date, no mechanism has been discovered to connect the disparate issues posed by robotics technologies and optimization methodologies. A smart in-house logistics system has the potential to improve the efficiency of the entire factory and production process.

8.4 Machine learning for robustness in advanced logistical planning

Examples of transportation-related issues where BD/ML can be used to identify patterns include vessel scheduling, bus scheduling as well as other transportation-related issues. A successful investigation could open up a new research area focused on robust optimization in logistics and transportation even though there are still technical challenges to overcome.

8.5 Last-mile delivery systems that use cutting-edge technology

The use of autonomous vehicles, augmented reality, and drones in last-mile delivery is one of the most hotly debated topics of the day. Since many new conceptual delivery models (like Mercedes' platoon and Amazon's flying warehouse) have been developed in recent years, new modeling approaches are expected to be created. According to our findings on the technical challenges of optimizing complexity, the effectiveness of solution approaches will play an important role in future research.

8.6 Outer space logistical considerations

To wrap things up, we'd like to talk about space travel. It is currently being developed and has already begun to use space transportation. The Starship spacecraft concept and prototype, as well as SpaceX4's Super Heavy rocket, have already undergone numerous feasibility studies. We'll be able to take our crew and cargo into space with these new, fully reusable space transport tools. Future deliveries are expected to be even more efficient thanks to Amazon's flying warehouse concept, which integrates space and air logistics.

References

[1] T. Parsons, The weight of cities: Urbanization effects on Earth's subsurface, AGU Adv. 2 (1) (2021). e2020AV000277.
[2] S.E. Bibri, Data-driven smart eco-cities of the future: an empirically informed integrated model for strategic sustainable urban development, World Futures (2021) 1–44.
[3] I.N. Greenberg, The interdependence of nature and nurture in the establishment and maintenance of mind: an eco-dynamic paradigm, 2021 (PhD discussion)., University College Cork.
[4] L. Aguiar-Castillo, V. Guerra, J. Rufo, J. Rabadan, R. Perez-Jimenez, Survey on optical wireless communications-based services applied to the tourism industry: potentials and challenges, Sensors 21 (18) (2021) 6282.
[5] M.D. Lytras, A. Visvizi, P.K. Chopdar, A. Sarirete, W. Alhalabi, Information management in smart cities: turning end users' views into multi-item scale development, validation, and policy-making recommendations, Int. J. Inf. Manag. 56 (2021), 102146.
[6] D. Settembre-Blundo, R. González-Sánchez, S. Medina-Salgado, F.E. García-Muiña, Flexibility and resilience in corporate decision making: a new sustainability-based risk management system in uncertain times, Glob. J. Flex. Syst. Manag. (2021) 1–26.
[7] I. Sevinc, B. Akyildiz, Reflection of digital transformation on cities: smart cities, Public Admin. Public Finance Res. 116 (2021).
[8] A.T. Salim, S.I. Basheer, N.R. Saadallah, A survey of using internet of things to enhance zero energy buildings, Texas J. Multidiscip. Stud. 1 (1) (2021) 205–213.
[9] S. Raja, The Do-Good Spirit: A Study in Unilever's Digital Communication of Corporate Social Responsibility (PhD diss.), Rutgers The State University of New Jersey, School of Graduate Studies, 2021.

[10] A. van der Hoogen, B. Scholtz, A.P. Calitz, Innovative digitalisation initiatives for smart communities and smart cities in a developing country, in: Resilience, Entrepreneurship and ICT, Springer, Cham, 2021, pp. 57–78.

[11] D. Enrico, A. Luigi, R. Damiano, D. Marilena, B.D. Agudo, G. Aldo, K. Tsvi, et al., Integrating citizen experiences in cultural heritage archives: requirements, state of the art, and challenges, J. Comput. Cultural Heritage (2021) 1–35.

[12] C. de Villiers, S. Kuruppu, D. Dissanayake, A (new) role for business–promoting the United Nations' sustainable development goals through the internet-of-things and blockchain technology, J. Bus. Res. 131 (2021) 598–609.

[13] S. Epting, The Morality of Urban Mobility: Technology and Philosophy of the City, Rowman & Littlefield, 2021.

[14] C.M.T. Castillo-Calzadilla, K. Zabala, P. Eneko Arrizabalaga, L. Hernandez, J.R. Mabe, J.M. Lopez, M.N. Casado, J.G. Santos, B. Molinete, The opportunity for smart city projects at municipal scale: implementing a positive energy district in Zorrozaurre, Ekonomiaz 4 (2021) 119–149.

[15] R. Basu, J. Ferreira, Sustainable mobility in auto-dominated Metro Boston: challenges and opportunities post-COVID-19, Transp. Policy 103 (2021) 197–210.

[16] F. Golbabaei, T. Yigitcanlar, J. Bunker, The role of shared autonomous vehicle systems in delivering smart urban mobility: a systematic review of the literature, Int. J. Sustain. Transp. 15 (10) (2021) 731–748.

[17] T. Litman, Autonomous Vehicle Implementation Predictions, Victoria Transport Policy Institute, Victoria, Canada, 2017.

[18] M.S. Whittingham, Materials challenges facing electrical energy storage, MRS Bull. 33 (4) (2008) 411–419.

[19] S.B. Sherman, Z.P. Cano, M. Fowler, Z. Chen, Range-extending zinc-air battery for electric vehicle, Aims Energy 6 (1) (2018) 121–145.

[20] V. Birss, E. El Sawy, S. Ketabi, P. Keyvanfar, X. Li, J. Young, Electrochemical energy production using fuel cell technologies, in: Handbook of Industrial Chemistry and Biotechnology, Springer, Cham, 2017, pp. 1729–1779.

[21] C. Suh, C. Fare, J.A. Warren, E.O. Pyzer-Knapp, Evolving the materials genome: how machine learning is fueling the next generation of materials discovery, Annu. Rev. Mater. Res. 50 (2020) 1–25.

[22] L. Kong, C. Yan, J.-.Q. Huang, M.-.Q. Zhao, M.-.M. Titirici, R. Xiang, Q. Zhang, A review of advanced energy materials for magnesium–sulfur batteries, Energy Environ. Mater. 1 (3) (2018) 100–112.

[23] D. Li, X. Huang, K. Han, C.-G. Zhan, Catalytic mechanism of cytochrome P450 for $5'$-hydroxylation of nicotine: fundamental reaction pathways and stereoselectivity, J. Am. Chem. Soc. 133 (19) (2011) 7416–7427.

[24] J.A. Keith, V. Vassilev-Galindo, B. Cheng, S. Chmiela, M. Gastegger, K.-R. Müller, A. Tkatchenko, Combining machine learning and computational chemistry for predictive insights into chemical systems, arXiv Preprint (2021). arXiv:2102.06321.

[25] R. Batra, L. Song, R. Ramprasad, Emerging materials intelligence ecosystems propelled by machine learning, Nat. Rev. Mater. (2020) 1–24.

[26] Y. Guan, D. Chaffart, G. Liu, Z. Tan, D. Zhang, Y. Wang, J. Li, L. Ricardez-Sandoval, Machine learning in solid heterogeneous catalysis: recent developments, challenges and perspectives, Chem. Eng. Sci. 117224 (2021).

[27] Y. Wang, K. Lopata, S.A. Chambers, N. Govind, P.V. Sushko, Optical absorption and band gap reduction in $(Fe_{1-x}Cr_x)_2O_3$ solid solutions: a first-principles study, J. Phys. Chem. C 117 (48) (2013) 25504–25512.

[28] O.A. von Lilienfeld, K.-R. Müller, A. Tkatchenko, Exploring chemical compound space with quantum-based machine learning, Nat. Rev. Chem. 4 (7) (2020) 347–358.

[29] Y. Qiu, Y. Lin, H. Yang, L. Wang, Ni-doped cobalt hexacyanoferrate microcubes as battery-type electrodes for aqueous electrolyte-based electrochemical supercapacitors, J. Alloys Compd. 806 (2019) 1315–1322.

[30] H. Kroll, G. Copani, E. Van de Velde, M. Simons, D. Horvat, A. Jäger, A. Wastyn, G.P. Abdollahian, M. Naumanen, An Analysis of Drivers, Barriers and Readiness Factors of EU Companies for Adopting Advanced Manufacturing Products and Technologies, World Scientific Press, 2016.

[31] L. Tianhong, L. Wenkai, Q. Zhenghan, Variations in ecosystem service value in response to land use changes in Shenzhen, Ecol. Econ. 69 (7) (2010) 1427–1435.

[32] R. Charan, S. Drotter, J. Noel, The Leadership Pipeline: How to Build the Leadership Powered Company, vol. 391, John Wiley & Sons, 2010.

[33] X. Chia, A.Y.S. Eng, A. Ambrosi, S.M. Tan, M. Pumera, Electrochemistry of nanostructured layered transition-metal dichalcogenides, Chem. Rev. 115 (21) (2015) 11941–11966.

[34] S.-Y. Chung, J.T. Bloking, Y.-M. Chiang, Electronically conductive phospho-olivines as lithium storage electrodes, Nat. Mater. 1 (2) (2002) 123–128.

[35] C.M. Lee, Changes in Sensory and Physicochemical Properties of Roasted Peanuts in Intermediate Moisture Foods (PhD diss.), University of Georgia, 2004.

[36] M. Alber, A.B. Tepole, W.R. Cannon, S. De, S. Dura-Bernal, K. Garikipati, G. Karniadakis, et al., Integrating machine learning and multiscale modeling—perspectives, challenges, and opportunities in the biological, biomedical, and behavioral sciences, NPJ Digital Med. 2 (1) (2019) 1–11.

[37] I. Sami, Z. Ullah, K. Salman, I. Hussain, S.M. Ali, B. Khan, C.A. Mehmood, U. Farid, A bidirectional interactive electric vehicles operation modes: vehicle-to-grid (v2g) and grid-to-vehicle (g2v) variations within smart grid, in: 2019 International Conference on Engineering and Emerging Technologies (ICEET), IEEE, 2019, pp. 1–6.

[38] S.M. Shariff, D. Iqbal, M.S. Alam, F. Ahmad, A state of the art review of electric vehicle to grid (V2G) technology, in: IOP Conference Series: Materials Science and Engineering, vol. 561 (1), IOP Publishing, 2019, p. 012103.

[39] S.A. Khan, R. Bohnsack, Influencing the disruptive potential of sustainable technologies through value proposition design: the case of vehicle-to-grid technology, J. Clean. Prod. 254 (2020), 120018.

[40] C. Bordons, F. Garcia-Torres, M.A. Ridao, Model Predictive Control of Microgrids, vol. 358, Springer, Cham, 2020.

[41] R. Fachrizal, M. Shepero, D. van der Meer, J. Munkhammar, J. Widén, Smart charging of electric vehicles considering photovoltaic power production and electricity consumption: a review, ETransportation 4 (2020) 100056.

[42] S. Behera, N.B. Dev Choudhury, A systematic review of energy management system based on various adaptive controllers with optimization algorithm on a smart microgrid, Int. Trans. Electr. Energy Syst. (2021), e13132.

[43] N. Wang, J. Li, S.-S. Ho, C. Qiu, Distributed machine learning for energy trading in electric distribution system of the future, Electr. J. 34 (1) (2021), 106883.

[44] J. Cao, D. Harrold, Z. Fan, T. Morstyn, D. Healey, K. Li, Deep reinforcement learning-based energy storage arbitrage with accurate lithium-ion battery degradation model, IEEE Trans. Smart Grid 11 (5) (2020) 4513–4521.

[45] M.T. Kahlen, W. Ketter, J. van Dalen, Electric vehicle virtual power plant dilemma: grid balancing versus customer mobility, Prod. Oper. Manag. 27 (11) (2018) 2054–2070.

[46] D. Ilić, S. Karnouskos, M. Beigl, Improving accuracy of energy forecasting through the presence of an electric vehicle fleet, Electr. Power Syst. Res. 120 (2015) 32–38.

[47] S.A.A. Kazmi, M.K. Shahzad, A.Z. Khan, D.R. Shin, Smart distribution networks: a review of modern distribution concepts from a planning perspective, Energies 10 (4) (2017) 501.

[48] D. Uckelmann, A definition approach to smart logistics, in: International Conference on Next Generation Wired/Wireless Networking, Springer, Berlin, Heidelberg, 2008, pp. 273–284.

[49] S. Liu, G. Zhang, L. Wang, IoT-enabled dynamic optimisation for sustainable reverse logistics, Proc. CIRP 69 (2018) 662–667.

[50] Y. Song, F. Richard Yu, L. Zhou, X. Yang, Z. He, Applications of the internet of things (IoT) in smart logistics: a comprehensive survey, IEEE Internet Things J. 8 (6) (2020) 4250–4274.

[51] M. Zöttl, J.G. Frommen, M. Taborsky, Group size adjustment to ecological demand in a cooperative breeder, Proc. R. Soc. B Biol. Sci. 280 (1756) (2013) 20122772.

[52] R. Baashirah, A. Abuzneid, SLEC: a novel serverless RFID authentication protocol based on elliptic curve cryptography, Electronics 8 (10) (2019) 1166.

[53] K. Buntak, M. Kovačić, M. Mutavdžija, Internet of things and smart warehouses as the future of logistics, Teh. Glas. 13 (3) (2019) 248–253.

[54] F.R. Batubara, J. Ubacht, M. Janssen, Challenges of blockchain technology adoption for e-government: a systematic literature review, in: Proceedings of the 19th Annual International Conference on Digital Government Research: Governance in the Data Age, 2018, pp. 1–9.

Further reading

[55] H. Marzbali, A.A. Massoomeh, M.J.M. Tilaki, M. Safizadeh, Moving the 2030 agenda ahead: exploring the role of multiple mediators toward perceived environment and social sustainability in residential neighbourhoods, Land 10 (10) (2021) 1079.

[56] A. Shamsuzzoha, J. Niemi, S. Piya, K. Rutledge, Smart city for sustainable environment: a comparison of participatory strategies from Helsinki, Singapore and London, Cities 114 (2021), 103194.

[57] C. Bıyık, A. Abareshi, A. Paz, R.A. Ruiz, R. Battarra, C.D.F. Rogers, C. Lizarraga, Smart mobility adoption: a review of the literature, J. Open Innov.: Technol. Market Complex. 7 (2) (2021) 146.

[58] E. Marlet, K. Carson, Clean Energy in the Bonneville Power Administration Area, 2021.

[59] J.A. Ondiviela, SmartCities. Technology as enabler, in: Beyond Smart Cities, Springer, Cham, 2021, pp. 103–185.

[60] H. Nevala, NGOs Fostering Alternative Tourism Economies: Sleeping Outdoors Campaign as a Case of Proximity Tourism, LUC kirjasto, 2021.

[61] C. Tennant, J. Stilgoe, The attachments of 'autonomous' vehicles, Soc. Stud. Sci. (2021). 03063127211038752.

[62] A. Fitzgerald, W. Proud, A. Kandemir, R.J. Murphy, D.A. Jesson, R.S. Trask, I. Hamerton, M.L. Longana, A life cycle engineering perspective on biocomposites as a solution for a sustainable recovery, Sustainability 13 (3) (2021) 1160.

[63] S. Bag, L.C. Wood, X. Lei, P. Dhamija, Y. Kayikci, Big data analytics as an operational excellence approach to enhance sustainable supply chain performance, Resour. Conserv. Recycl. 153 (2020), 104559.

[64] A. De La Vega-Leinert, J.K. Cristina, M. Jiménez-Moreno, C. Steinhäuser, Young people's visions for life in the countryside in Latin America, Geogr. Rev. (2021) 1–25.

[65] O.S. Neffati, S. Sengan, K.D. Thangavelu, S.D. Kumar, R. Setiawan, M. Elangovan, D. Mani, P. Velayutham, Migrating from traditional grid to smart grid in smart cities promoted in developing country, Sustain. Energy Technol. Assess. 45 (2021) 101125.

[66] A.E. Omolara, A. Alabdulatif, O.I. Abiodun, M. Alawida, A. Alabdulatif, H. Arshad, The internet of things security: a survey encompassing unexplored areas and new insights, Comput. Secur. (2021) 102494.

[67] E. Arrilucea, M.N. Bilbao, J. Herrera, J. Del Ser (Eds.), Innovation policies and big data: opportunities and challenges, in: Perspectives for Digital Social Innovation to Reshape the European Welfare Systems, IOS Press, Washington, DC, 2021, pp. 159–181.

[68] L.H. Saw, Y. Ye, A.A.O. Tay, Integration issues of lithium-ion battery into electric vehicles battery pack, J. Clean. Prod. 113 (2016) 1032–1045.

[69] Y. Jin, K. Liu, J. Lang, D. Zhuo, Z. Huang, C.-a. Wang, W. Hui, Y. Cui, An intermediate temperature garnet-type solid electrolyte-based molten lithium battery for grid energy storage, Nat. Energy 3 (9) (2018) 732–738.

[70] A. Vetrò, L. Canova, M. Torchiano, C.O. Minotas, R. Iemma, F. Morando, Open data quality measurement framework: definition and application to open government data, Govt. Inf. Q. 33 (2) (2016) 325–337.

[71] J. Yoon, H.-.S. Yang, B.-.S. Lee, W.-.R. Yu, Recent progress in coaxial electrospinning: new parameters, various structures, and wide applications, Adv. Mater. 30 (42) (2018) 1704765.

[72] J. Chong, O. Soufan, C. Li, I. Caraus, S. Li, G. Bourque, D.S. Wishart, J. Xia, MetaboAnalyst 4.0: towards more transparent and integrative metabolomics analysis, Nucleic Acids Res. 46 (W1) (2018) W486–W494.

[73] V.J. Hodge, S.M. Devlin, N.J. Sephton, F.O. Block, P.I. Cowling, A. Drachen, Win prediction in multi-player esports: live professional match prediction, IEEE Trans. Games 13 (2019) 368–379.

[74] A.W. Cox, Fidelity Assessment for Model Selection (FAMS): A Framework for Initial Comparison of Multifidelity Modeling Options (PhD diss.), Georgia Institute of Technology, 2019.

[75] Notations, Frequently Used. Joint Load Scheduling and Voltage Regulation in Distribution System with Renewable Generators.

[76] C. Latinopoulos, Efficient Operation of Recharging Infrastructure for the Accommodation of Electric Vehicles: A Demand Driven Approach, Centre for Transport Studies, Department of Civil and Environmental Engineering, Imperial College London, 2015.

[77] W. Ding, Study of smart warehouse management system based on the IOT, in: Intelligence Computation and Evolutionary Computation, Springer, Berlin, Heidelberg, 2013, pp. 203–207.

[78] Y. Zhou, S. Zheng, G. Zhang, Artificial neural network based multivariable optimization of a hybrid system integrated with phase change materials, active cooling and hybrid ventilations, Energy Convers. Manag. 197 (2019), 111859.

PART SIX

Tackling cyber attacks

CHAPTER TWENTY

Generative adversarial network-based deep learning technique for smart grid data security

Varsha Himthani[a] and Vivek Prakash[b]
[a]Manipal University Jaipur, Jaipur, Rajasthan, India
[b]Department of Electrical and Electronics Engineering, Banasthali Vidyapith, School of Automation, Tonk, India

1. Introduction

The global policy of clean energy utilization strongly encourages the creation of smart grids to expedite the major generation share from renewable energy sources (RES). The smart grid (SG) is an intelligent electricity grid that implements evolving information and communication infrastructure for controlling power generation, transmission, distribution, and consumers [1]. In contrast to traditional electricity grids, SG facilitates a two-way dataflow between the consumer and electrical utilities as shown in Fig. 1 [2]. This facility enables real-time electricity generation based on consumers' energy demands. Hence, consumer data secrecy becomes imperative while energy usage data collection [3]. There are multiple inevitable security challenges that need to be addressed at the levels of consumer, communication, and energy provider. SG data security facets such as privacy, validation, approval, etc., require in-depth investigation to implement data security. The study of SG data security has gained wide attention among the global researchers and industries working in this area [4]. The main SG data privacy concerns are as follows:

- Data ownership
- Data access
- Regulation on customer data use
- Data security from criminal activity or risk of surveillance
- Use or transfer of data for commercial purpose
- Identity disclosure
- Reveal personal usages patterns
- Specific appliances usage determination

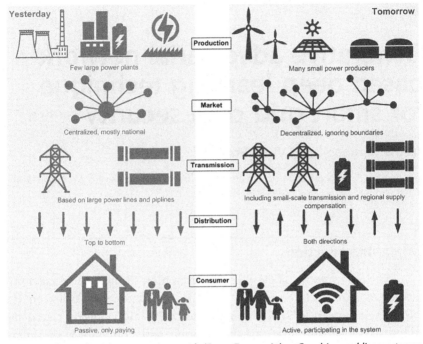

Fig. 1 Traditional grid versus smart grid. *(From Energy Atlas: Graphics and license terms, n.d.)*

In SG, smart meters would monitor customer electricity consumption. The data from smart meters may expose usage data of residential consumers and actions within the house as shown in Fig. 2 [5].

The electrical utilities or load-serving entities may use the consumer data to fulfil their commercial interest. In the current competitive market, specific user data could be used by some specific electricity providers for unfair commercial gains [1,6]. Appropriate data encryption techniques could be a crucial solution for data protection from unfair usages. Hence, these concerns necessitate a robust data security technique for both consumers and the electrical utilities [7,8]. In this regard, this work develops a novel deep learning technique for SG data security. Data encryption and data embedding are two ways of protecting data from unauthorized access. The traditional embedding techniques are easily detectable by steganalysis tools. Nowadays, embedding techniques based on deep learning are emerging as a potential solution that provides high security and high payload capacity [3]. In the proposed work, the combination of encryption and embedding is

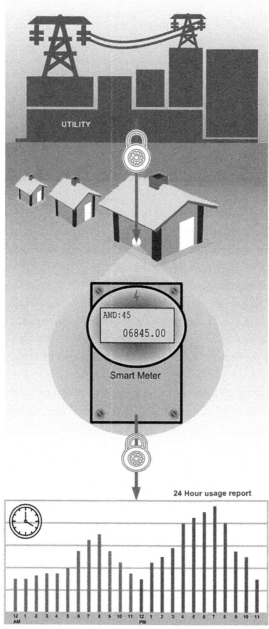

Fig. 2 Consumer electricity consumption data protection. *(From Data Privacy. (2021). https://www.whatissmartenergy.org/data-privacy. Pictorial depiction of Data consumption, smart metering, data privacy and communication between SG and consumer.)*

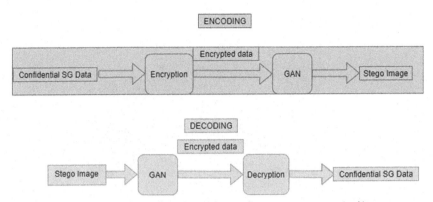

Fig. 3 Workflow of GAN-based SG data security. *(No permission required.)*

used to provide high security. Data are encrypted first and then the encrypted data are embedded using generative adversarial network (GAN). The encrypted data are protected in cover image produced by GAN. This technique is widely used in image generation but not used significantly with embedding as shown in Fig. 3.

GANs have the ability to produce good quality stego images with minutest distortion. Hence, attacker would not be able to recognize that confidential information is hidden in the cover image. In this method, the confidential information is encrypted; therefore, this method could provide higher security.

In the remaining chapter, Section 2 details out the problem formulation. Section 3 discusses the various stages of proposed methodology. Section 4 provides the details of data used, results, and outcomes. Section 5 draws the conclusive remarks.

2. Problem formulation and design considerations

This section details out the mathematical modeling used to encrypt the data. In this work, one of the benchmark cryptography techniques is used for the data encryption, widely known as chaotic maps [9].

2.1 Chaotic encryption

The chaotic maps provide excellent random behavior that provides highly secure keys with high key space and sensitivity [9]. In the proposed method, 3D chaotic maps are used to encrypt the input data image [10]. The system iterated three chaotic sequences in 3D chaotic maps as given in Eqs. (1), (2) and (3):

$$x = Ax - Ax^2 + Bxy^2 + Cz^3 \tag{1}$$
$$y = Ay - Ay^2 + Byz^2 + Cx^3 \tag{2}$$
$$z = Az - Az^2 + Bzx^2 + Cy^3 \tag{3}$$

Eqs. (1), (2), and (3) generate three random sequences:

$$X = x_0, x_1, \ldots x_n$$
$$Y = y_0, y_1, \ldots y_n$$
$$Z = z_0, z_1, \ldots z_n$$

The initial values of x_0, y_0, and z_0 should be taken between 0 and 1. There are three constants A, B, and C used in equations that make more secure chaotic map [11].

2.2 GAN

In GANs, the generative models are designed through convolutional neural networks (CNN) [12]. These models are based on unsupervised machine learning that learns automatically from the input data patterns [13]. The model can be used to develop or output new sample data from the original data [14]. There are two submodels of GANs:

a) Generator: This model is trained to generate new samples.

b) Discriminator: This model classifies between the real and sample (i.e., generated) data.

The generator and discriminator both are trained together until the discriminator correctly classify the samples created by the generator, which means the generator would create the cogent samples [15].

With the recent advancement in GANs, the data-hiding through GAN is widely implemented because of the properties like imperceptibility and high-quality reconstruction [16]. There are two networks designed to hide data through GANs, called encoder and decoder networks. The encoder network is composed of a generator and discriminator [17,18]. The generator is used to learn data-hiding in the cover image, and the discriminator compares the original cover image with the stego image (generated by the generator). The decoder network is used to extract the secret image from a stego image. The decoder is parallelly trained with the encoder network [19].

3. Proposed methodology

This section presents the various stages used for the implementation of the proposed data security model. In the proposed method, there are two stages. First, the confidential data are encrypted by 3D chaotic encryption and converted into unreadable form [10]. Second, the encrypted data are implanted into a cover image using GAN to provide advanced security [20]. Fig. 4 shows the block diagram of the proposed methodology.

3.1 Data encryption

The confidential data are taken in the form of graphical representation of data in 128*128 size gray scale image. This confidential image is encrypted using a 3D chaotic map [11]. The chaotic maps generate three random sequences that would be used as secret keys for encryption and image pixels are permuted based on these random sequences. In Eqs. (1), (2), and (3), the initial values of x_0, y_0, and z_0 are taken as 0.24, 0.35, and 0.74, respectively, to get the random sequence. The values of constants A, B, and C are taken as 3.6, 0.02, and 0.01, respectively [10]. After permutation, we get the encrypted image that is noise-like.

3.2 Data-hiding through GAN

The encrypted image is embedded in the cover through a generator of the encoder network. The generator is trained to hide the confidential data into a cover image [21]. Fig. 5 depicts the architecture of generator. It consists of four convolutional and deconvolutional layers. In addition to that, nine layers of residual blocks are employed in between convolution and deconvolution layers. Each layer is followed by rectified linear unit (ReLU) activation function [22] and instance normalization except the last layer, which is followed by a Tanh activation function [20].

Moreover, the discriminator is used to distinguish the stego image from cover images to generate the high-quality stego image [23]. Fig. 6 depicts the architecture of discriminator. It consists of six convolution layers and one fully connected soft max layer. Leaky ReLU activation is employed as activation [20]. The generator is trained till it generates the realistic image that discriminator would not be able to distinguish it from original cover images [19].

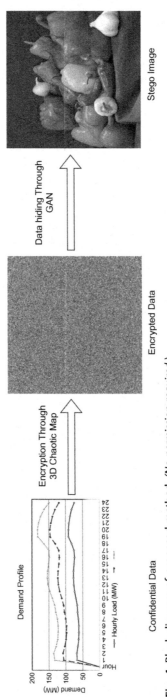

Fig. 4 Block diagram of proposed method. *(No permission required.)*

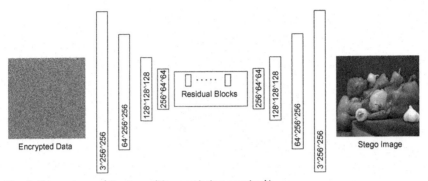

Fig. 5 Generator architecture. *(No permission required.)*

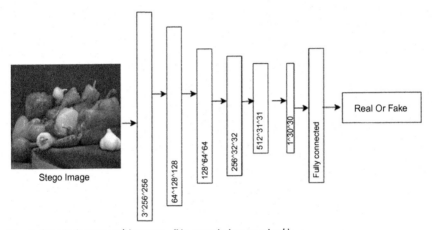

Fig. 6 Discriminator architecture. *(No permission required.)*

In GANs, the generator tries to minimize and discriminator tries to maximize the following loss function

$$\text{Loss function} = E_R[\log(D(x))] + E_F[\log(1 - D(G(F)))] \quad (4)$$

E_R and E_F are the expected values for real and fake data, respectively. D is the discriminator estimated values for real data; x and $G(F)$ are the fake data instance [24].

4. Result and discussion

4.1 Experimental setup

The data sets used for training purpose is available online [25]. The data set used as cover images consists of 900 color images and 900 gray-scale images as secret images. The size of these images is taken as 128×128. To carry out the training, the batch size of 8 is selected and the learning rate is set to 0.0001. Further, the test data set comprising 50 encrypted images that are used for the evaluation of the model performance. Fig. 7 depicts the results of some test images. In this, the first column shows the original data images, second column represents the encrypted data images, third column illustrate the stego images, and the last column represents the regenerated images after decoding.

The visual quality of stego and regenerated image is evaluated on the basis of peak signal-to-noise ratio (PSNR) and structural similarity index (SSIM). These parameters are used to measure the difference in image quality between two images, as the pixel permutation and embedding process will degrade the image quality [26]. The PSNR is calculated as

$$PSNR = 10 \log_{10}\left(\frac{d^2}{MSE}\right) \tag{5}$$

Here, d represents the maximum difference between pixels of two comparable images and MSE is the mean square error, which can be calculated as following:

$$MSE = \frac{\sum_{a=1}^{M}\sum_{b=1}^{N}[CI(a,\ b) - Ri(a,\ b)]^2}{M \times N} \tag{6}$$

In the above equation, $M \times N$ is the size of the image, CI and Ri represent the confidential data image and regenerated image, a and b are the coordinates of image pixels. Lesser the mean square error and the higher the PSNR value. Therefore, the higher PSNR shows the good-quality image [27,28].

SSIM evaluation of structural similarity between two comparable images can be calculated as

$$SSIM(I_1, I_2) = \frac{(2I_1 I_2 + C_1)(2\sigma_1\sigma_2 + C_2)}{(I_1^2 + I_2^2 + C_1)(\sigma_1^2 + \sigma_2^2 + C_2)} \tag{7}$$

where I_1 and I_2 are the two comparable images, σ_1 and σ_2 are the standard deviation of images, and C_1 and C_2 are the constants.

Table 1 depicts the PSNR and SSIM values of images. The first row shows the PSNR and SSIM values that are calculated between original cover and stego image. It can be observed that PSNR value is 40.04 dB and 99% structural similarity that shows the high visual quality of stego image.

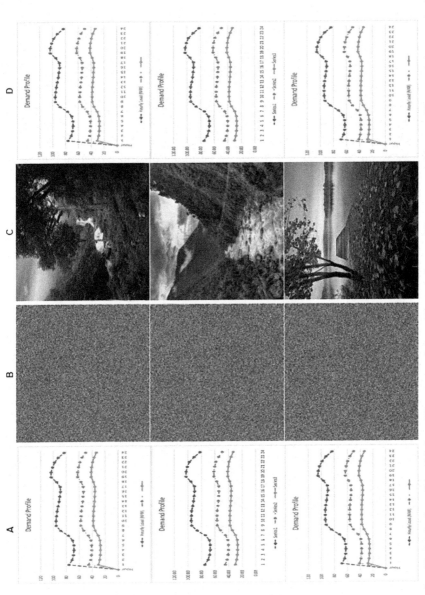

Fig. 7 Results. (A) Original confidential data, (B) encrypted data, (C) stego image, (D) regenerated image. *(No permission required.)*

Table 1 PSNR and SSIM image quality parameters.

	PSNR (Decibel)	SSIM
Original cover and stego image	40.04	0.9903
Original confidential data and regenerated image	39.01	0.9889

Further, it could be inferred from the obtained PSNR and SSIM parameters that the transmitted image is imperceptible. The second row depicts the PSNR value 39.01 dB and 98% structural similarity that are calculated between original confidential data image and the regenerated image. Hence, the regenerated image at the receiver's end also perceives good visual quality.

5. Conclusion

This chapter presents a novel GAN-based deep learning technique for SG data security. The customer electricity consumption data are secured using the GAN-based data encryption and data-hiding technique. The PSNR and SSIM values are estimated to check the visual quality and robustness of the stego image. It is analyzed from the obtained parameter that the visual quality of the stego image is high and the transmitted image is imperceptible from the unauthorized data intrusion. The outcomes of the proposed methodology would not only be helpful to securely manage the bidirectional communication between the SG and consumers but also encourage consumers to actively participate in the grid management activities without losing the data privacy. This work could be potentially enhanced by implementing highly efficient encryption techniques like 7D chaotic map.

References

[1] H. Sun, N.D. Hatziargyriou, H.V. Poor, L. Carpanini, M.A.S. Fornié (Eds.), Smarter Energy: From Smart Metering to the Smart Grid, IET, 2016, https://doi.org/10.1049/PBPO088E.

[2] Energy Atlas (n.d.) Energy Atlas: Graphics and license terms.

[3] S. Sengan, V. Subramaniyaswamy, V. Indragandhi, P. Velayutham, L. Ravi, Detection of false data cyber-attacks for the assessment of security in smart grid using deep learning, Comput. Electr. Eng. 93 (2021), 107211, https://doi.org/10.1016/j.compeleceng.2021.107211.

[4] J. Tian, H. Li, R. Chen, The emerging of smart citizen concept under smart city environment, in: Proceedings of the International Conference on Electronic Business (ICEB), vol. 2018, International Consortium for Electronic Business, 2018, pp. 739–742. http://iceb.nccu.edu.tw/.

[5] Data Privacy, 2021. https://www.whatissmartenergy.org/data-privacy.

[6] A.K. Das, S. Zeadally, Chapter 13—Data Security in the Smart Grid Environment, Elsevier BV, 2019, pp. 371–395, https://doi.org/10.1016/b978-0-08-102592-5.00013-2.

[7] B.P. Bhattarai, S. Paudyal, Y. Luo, M. Mohanpurkar, K. Cheung, R. Tonkoski, R. Hovsapian, K.S. Myers, R. Zhang, P. Zhao, M. Manic, S. Zhang, X. Zhang, Big data analytics in smart grids: state-of-the-art, challenges, opportunities, and future directions, IET Smart Grid 2 (2) (2019) 141–154, https://doi.org/10.1049/iet-stg.2018.0261.

[8] S. Zeadally, A.S.K. Pathan, C. Alcaraz, M. Badra, Towards privacy protection in smart grid, Wireless Personal Commun. 73 (1) (2013) 23–50, https://doi.org/10.1007/s11277-012-0939-1.

[9] Abraham, N. Daniel, Secure image encryption algorithms: a review, Int. J. Sci. Technol. Res. 2 (2013) 186–189.

[10] M.B. Hossain, M.T. Rahman, A.B.M.S. Rahman, S. Islam, A new approach of image encryption using 3D chaotic map to enhance security of multimedia component, in: 2014 International Conference on Informatics, Electronics and Vision, ICIEV 2014, IEEE Computer Society, 2014, https://doi.org/10.1109/ICIEV.2014.6850856.

[11] A. Kanso, M. Ghebleh, An algorithm for encryption of secret images into meaningful images, Opt. Lasers Eng. 90 (2017) 196–208, https://doi.org/10.1016/j.optlaseng.2016.10.009.

[12] K. Wang, C. Gou, Y. Duan, Y. Lin, X. Zheng, F.Y. Wang, Generative adversarial networks: Introduction and outlook, IEEE/CAA J. Autom. Sin. 4 (4) (2017) 588–598, https://doi.org/10.1109/JAS.2017.7510583.

[13] H. Shi, J. Dong, W. Wang, Y. Qian, X. Zhang, SSGAN: Secure steganography based on generative adversarial networks, in: Lecture Notes in Computer Science (including subseries Lecture Notes in Artificial Intelligence and Lecture Notes in Bioinformatics), vol. 10735, Springer Verlag, 2018, pp. 534–544, https://doi.org/10.1007/978-3-319-77380-3_51.

[14] W. Tang, S. Tan, B. Li, J. Huang, Automatic steganographic distortion learning using a generative adversarial network, IEEE Signal Process. Lett. 24 (10) (2017) 1547–1551, https://doi.org/10.1109/LSP.2017.2745572.

[15] D. Volkhonskiy, I. Nazarov, E. Burnaev, Steganographic generative adversarial networks, in: Proc. SPIE, 11433, 2020, https://doi.org/10.1117/12.2559429.

[16] X. Duan, N. Liu, M. Gou, W. Wang, C. Qin, StegoCNN: Image steganography with generalization ability based on convolutional neural network, Entropy 22 (10) (2020) 1–15, https://doi.org/10.3390/e22101140.

[17] R. Zhang, S. Dong, J. Liu, Invisible steganography via generative adversarial networks, Multimed. Tools Appl. 78 (7) (2019) 8559–8575, https://doi.org/10.1007/s11042-018-6951-z.

[18] J. Yang, D. Ruan, J. Huang, X. Kang, Y.Q. Shi, An embedding cost learning framework using GAN, IEEE Trans. Inform. Foren. Security 15 (2020) 839–851, https://doi.org/10.1109/TIFS.2019.2922229.

[19] F. Li, Z. Yu, C. Qin, GAN-based spatial image steganography with cross feedback mechanism, Signal Process. 190 (2022), https://doi.org/10.1016/j.sigpro.2021.108341.

[20] X. Liu, Z. Ma, X. Guo, J. Hou, G. Schaefer, L. Wang, V. Wang, H. Fang, Camouflage generative adversarial network: coverless full-image-to-image hiding, in: IEEE Transactions on Systems, Man, and Cybernetics: Systems, vols. 2020, Institute of Electrical and Electronics Engineers Inc., 2020, pp. 166–172, https://doi.org/10.1109/SMC42975.2020.9283054.

[21] D. Hu, L. Wang, W. Jiang, S. Zheng, B. Li, A novel image steganography method via deep convolutional generative adversarial networks, IEEE Access 6 (2018) 38303–38314, https://doi.org/10.1109/ACCESS.2018.2852771.

[22] H. Ide, T. Kurita, Improvement of learning for CNN with ReLU activation by sparse regularization, in: Proceedings of the International Joint Conference on Neural Networks, vol. 2017, Institute of Electrical and Electronics Engineers Inc, 2017, pp. 2684–2691, https://doi.org/10.1109/IJCNN.2017.7966185.

[23] G. Xu, H.Z. Wu, Y.Q. Shi, Structural design of convolutional neural networks for steganalysis, IEEE Signal Process. Lett. 23 (5) (2016) 708–712, https://doi.org/10.1109/LSP.2016.2548421.

[24] Developers (n.d.). https://developers.google.com/machine-learning/gan/loss.

[25] Natural Scenes. (n.d.). Available online http://natural-scenes.cps.utexas.edu/db.shtml.

[26] U. Sara, M. Akter, M.S. Uddin, Image Quality Assessment through FSIM, SSIM, MSE and PSNR—A Comparative Study, J. Comput. Commun. 8–18 (2019), https://doi.org/10.4236/jcc.2019.73002.

[27] A. Horé, D. Ziou, Image quality metrics: PSNR vs. SSIM, in: Proceedings of the International Conference on Pattern Recognition, 2010, pp. 2366–2369, https://doi.org/10.1109/ICPR.2010.579.

[28] Q. Li, X. Wang, X. Wang, B. Ma, C. Wang, Y. Xian, Y. Shi, A novel grayscale image steganography scheme based on chaos encryption and generative adversarial networks, IEEE Access 8 (2020) 168166–168176, https://doi.org/10.1109/ACCESS.2020.3021103.

PART SEVEN

Smart communications

CHAPTER TWENTY-ONE

An overview of smart city planning—The future technology

Swetha Shekarappa G.[a], Manjulata Badi[a], Saurav Raj[b], and Sheila Mahapatra[a]

[a]Department of Electrical and Electronics Engineering, Alliance University, Bangalore, Karnataka, India
[b]Department of Electrical Engineering, Institute of Chemical Technology, Jalna, India

1. Introduction

The term "smart city" refers to a framework that combines existing infrastructure with modern information and networking technology to provide a full set of productive urban services [1]. A smart city is one in which basic infrastructure, Internet technology connectivity, civic amenities, and economic connectivity are all linked in order to improve the area's aggregate awareness [2]. Smart cities are massive, complex, and reliant on technology, and they confront a slew of technological, financial, social, and ethical concerns [3]. Economical and financial concerns, individuals' ever-changing needs, community collaboration, consumer adhesive bonding, stability, and protection are just a few of the concerns and obstacles that smart cities confront [4]. Clever governance, highly educated people, smart finance, wise transit, ecology, and dwelling are the six dimensions in which cities can benefit from being smart [5]. By delivering suitable and accessible facilities, smart cities met the needs of corporations, individuals, and organizations. Environmental, mobility, medical, tourist, energy conservation, and residential security are just a few of the areas where metropolitan operations can be enhanced [6]. Considering the merits of smart city development for inhabitants, organizations, the ecology, and other factors, these societies are vulnerable to a variety of cyber risks, making it difficult to achieve the method approach [7]. A single or institution's susceptible activity in a smart city can put the overall community at threat [8]. This sprawling metropolis also presents a significant problem for digital court operations. In a smart city, securing cybersecurity entails protecting data and the mesh from assaults and criminal behavior [9]. Suppliers rarely analyze the cybersecurity of software and hardware designed for smart cities. As a result, the adoption of such insecure items can result in the database being flooded with phony data, the network being shut down, and the network malfunctioning because of hacking [10]. Aside from cybersecurity, there's also the question of confidentiality and people's interactions with the administration [11]. Customers' utilization of these technologies will be questioned due to the likelihood of invasion into consumers' confidentiality and the absence of cybersecurity in smart urban [12]. Identifying cybersecurity issues and risks to individuals' information is the first key to tackling cybersecurity concerns in smart cities and preserving community members' security [13]. It is impossible to build, execute, and grow a smart city instead of being conscious of these issues and giving effective responses [14]. The country's population, on either extreme, is expected to exceed 7.9 billion people by the end of 2050, as per UN data. Around 75% of the worldwide people will be concentrated in cities, with numerous cities having populations exceeding 10 million inhabitants.

Deep learning is being commonly employed on data gathered by academics. Deep learning is a way of artificial intelligence and machine learning concepts, which playing major roles for solving problems in various fields of research. The main goal of AI is to create intelligent solutions and to imbue machines with human intelligence, and it has a wide range of implications in smart cities [15]. Deep learning can collect and examine data in real time, allowing the system to react to changing environments [16]. Artificial intelligence is a technique for developing systems, taking decisions, problem-solving, perception, learning and linguistic intelligence, and mimic human behavior. Electrical and computer engineers contribute to the planning, creation, evaluation, and production procedures for younger generations of gadgets and technology, putting them at the frontline of intellectual invention. Though these experts aspire for advancement, their goals may collide with artificial intelligence's constantly rising implications. Applications of AI include health care, automotive industry, robotics, agriculture, e-commerce, education, data security, and social media. Increased competition, productivity, a scarcity of statistics, essential for appropriate management, shifting supply, and demand trends are all posing problems to the smart technology. Deep learning is a subfield of artificial intelligence, which aids machine learning by utilizing neural networks. Machine learning is a subsection of artificial intelligence, and this uses different types of data to solve tasks. Object recognition programs have achieved considerable gains in computational linguistics, automation, and many other fields in recent decades compared with standard machine learning approaches [17].

Machine learning algorithms learn user data first, and then, it automatically solves predictive tasks.

There are three main important kinds of machine learning techniques like supervised unsupervised, and reinforcement learning. Experts in machine learning and electrical engineering use AI to create and optimize technologies, as well as give latest data inputs for AI to analyze.

It is critical to understand the city government's functions and obligations. Irrespective of the goal, one of these tasks is to create a healthy and sustainable city for the inhabitants by putting well-being at the center of the city. As civilization and industrialization rise at a rapid pace, most major cities have endured from poor air quality. Air quality is a growing source of concern among government agencies and citizens, as it has an impact on several aspects of the human environment and progress. In fact, the subject of how municipal governments adapt, particularly considering global and technological advances, necessitates a thorough examination. Crucial concerns such as transport, health care, education, economy, and the ecology will be studied in this perspective.

The usage of technologies that can assist in the event of critical situations is an important aspect of the framework of smart transportation systems (STSs) in green infrastructure. Natural disasters such as tremors, hurricanes, and storms, as well as man-made disasters such as terrorist acts and toxic sludge leaks, are manifestations of critical situations. Disasters are frequently accompanied by the devastation of local network capacity, if any exists, posing serious challenges for rescue operations.

Robust system modularity, the accessibility of open source, effective connectivity, and the flexibility of technological solutions are all critical for emergencies and protection missions. Any blend of multiple smart city national infrastructure and services can be adopted, relying on the commercial grade, financing accessibility, and sophistication of innovations. The incorporation of new technological advancements infrastructure and transport systems increased civic engagement in the creation—implementation—operations of such systems, vibrant leadership, and effective resource allocation are all important factors in the viability of smart city operations.

The intricate cyber-physical systems (CPS) that are the topic of this book are woven together by thousands of networked sensors, baseline connectivity, and analytics platforms. The smart city vision is examined holistically from the perspective of complex adaptive systems in this chapter. This paper conducts an extensive literature review on developing and forecasting new smart cities in the face of significant deployment to present a holistic concept of smartness and its scope at various spatial scales. The characteristics of several emergent smart city services and systems are then addressed. According to the study, a clever mix of technology-based and nontechnology-based remedies not only improves metropolitan performance but also improves residents' quality of life.

The focus of this chapter is to explore and identify an overview of the smart city planning, which is the upcoming future technology. The initial section incorporates the introduction, which is to come up with realistic and viable ways to alleviate or mitigate these issues. Other portions of the current work are structured as, in the second section, after introducing the smart city planning, it deals with the AI, ML, and deep learning in smart cities. In the third section, the overview of smart city planning with respect to various subsections is considered. The details of the cybersecurity in smart cities are provided in the fourth section. In the fifth section, deep learning for cybersecurity and user privacy in smart cities is explained. Finally, the last section is devoted to the conclusion.

2. Approach to artificial intelligence, machine learning, and deep learning for smart city planning

According to the United Nations Department of Economic and Social Affairs, urban regions presently house 55% of the earth's population, which is anticipated to climb to 68% by 2050. Indian cities account for about 31% of the country's present populations and provide 63% of GDP. By 2030, cities will house 40% of India's population and provide 75% of the country's GDP. With such rapid urbanization, upgrading cities' structural, organizational, sociological, and financial facilities is critical. Algorithmic technologies based on artificial intelligence (AI), machine learning (ML), and deep learning are used to do this.

Deep learning is a subset of machine learning that focuses on the research and implementation of learn systems [18]. In alternative aspects, deep learning with data acquisition, analogous to a brain, seeks to retrieve certain relevant features, despite the quantity of consecutive stages in its framework, in addition to construct a prototype for problem-solving judgment [19]. Deep learning can exploit the occurrence of these many multiple levels to uncover key characteristics of the challenge in each level and use them to make smart judgments in addressing the difficulty [20]. Deep learning is predicated on the continuous identification and exploration of complicated data structures.

The learning mechanism is accomplished by creating neural networks, which are data structures influenced by the organization of the neuron [21]. The network's architecture is made up of numerous operational levels. Deep learning looks for the efficacy of an appropriate experimental in the input method, in order to find a decent depiction using a sequential pattern of conceptions similar to the processing elements [22]. Upon understanding from unlabeled input information, deep learning may now generate new statistics. As a result, it has earned the moniker of "creative intelligence." It's worth noting that deep learning, which is based on artificial intelligence, has joined the area. Artificial intelligence has benefited from this understanding to adapt more intuitively to human desires.

Artificial intelligence and machine learning algorithms are now fundamental parts of industries and hence are crucial to building smart cities. The deployment of AI-enabled intelligent machines powered by machine learning creates cyber-physical space, which includes traffic sensors, medical monitoring, industrial control systems, video cameras, environment sensors, and smart meters. The data obtained through all these intelligent machines when analyzed help in predictive analysis and decision-making for smart city planning. From preserving a safer world to improving public transportation and security, intelligence and machine learning algorithms have many applications in a smart city. A city can develop for improved traffic signal strategies by combining AI and machine learning strategies with IoT to ensure that residents get through one location to other as easily and securely as feasible. Machine learning gathers data via a variety of sources and sends it to a centralized database for subsequent processing. Once the documentation is acquired, it must be used to make a city smarter.

Machine learning and AI, on either extreme, can aid with collecting waste, administration, and clearance, which is an important municipality function in each city. As a result, efficient composting and trash managing technologies allows for a long-term solid waste management. Artificial intelligence (AI) has the capacity to comprehend how communities are utilized and operate. It aids urban planners in gaining a better understanding of whether the city reacts to numerous changes and projects.

In this approach, AI-powered object detection algorithms might, for example, empower machines to recognize millions of aspects of metropolitan life in a symphony, such as individuals, government employees, vehicles, incidents, wildfires, catastrophes, rubbish, and plenty more. Not only does the program enables for unsupervised surveillance, but it also enables for choices to be taken depending on the success of each of these pieces, their evolving behaviors over the period of the day or time, and their interactions to smart city applications.

While AI and machine learning are revolutionizing the manner municipalities manage, distribute, and maintaining national services, they are not without flaws. As a result, adapted alternatives must be considered in order to keep urban planning efforts going. As a result, contemporary smart city projects based on AI and machine learning appear to improve city services and lifestyles in sectors such as transport, illumination, protection, connection, and universal health care, among many others.

3. Overview of smart city

Urban areas have become better and greater technologically advanced in current history. Cities can create excellent usage their resources, conserve revenue, and provide superior benefits to their inhabitants by combining emerging innovations with rapid and accessible connectivity.[23]. Cities are paying increasing emphasis to offering a high standard of living and a competitive commercial condition in order to retain money, new immigrants, and tourism [24]. Government has decided that, though restricted funds, inadequate resources, and out-of-date processes usually impede their objectives, emerging solutions can transform such obstacles into possibilities [25]. A smart city, pertaining to [26], is one that uses technology to automate and change governmental functions in order to better inhabitants' lives. Smart cities use digital technology to improve the efficiency and accessibility of municipal infrastructure, reduce monetary load and project provides, and engage with inhabitants more proactively and productively [27]. Fields such as public services and congestion, transit, electricity, irrigation, health care, and waste disposal have evolved in response of smart city technologies [28,29]. The Internet of things (IoT) detectors used in the city of the future are of several forms. Parking guidance devices, physical properties monitoring, real-time urban noise monitoring, adaptive cruise control, lanes optimizing, and lighting controls are examples of these sensing devices [30]. The Internet of things (IoT) is a live innovation that is employed in the smart urban elements stated above. The Internet, on the other hand, is a live tool for collecting and analyzing centralized smart city statistics [31]. The major components that can be included in smart city planning have been discussed below.

3.1 Smart energy

Conventional electricity grid architecture is insufficient to fulfill the demands of increasing population. A contemporary sophisticated electricity system is required to meet the demands for dependability, adaptability, management, renewable and clean energy generation, and cost-effectiveness [32]. A smart power grid that uses IC modern technologies can support two-way interaction and electric fields among various geometrical components [33]. The smart grid allows for real-time surveillance, guaranteeing that power exchanges between the electricity network and consumers are as efficient as possible. By incorporating renewable sources into the network, it also allows for the generation of eco-friendly electricity on both the power company and consumers [29,34,35].

3.2 Smart health care

Smart health care is a term of medical system that connects consumers, medical facilities, and organizations by utilizing technology such as wearables, the Internet of things, and mobile broadband to proactively obtain data. Ultimately, it actively manages the environment's demands and reacts to them more intelligently [36]. Specialists, patient, clinics, and pharmaceutical research facilities are the primary ingredients of smart health care. Disease control, clinical services, diagnostic tests, healthcare administrators, medical decision-making process, and medical science are all aspects of cognitive healthcare coverage. Electronically linking technological devices to healthcare hubs and data processing platforms can enable remote management [37]. Fig. 1 depicts the procedure in a smart city usage of digital health care.

3.3 Smart transportation

The effective utilization of preexisting infrastructure and current technology is one of the strategists' priorities in contemporary transportation infrastructure. Several of the implement and test of traffic management systems in this respect is to improve system effectiveness while also increasing automotive and consumer health and reducing commute time. To attain the aforementioned aim, the transport system requires optimal technologies to service the automotive industry, as well as competent management services [38]. The most significant benefits of implementing smart transportation systems include reduced road congestion, increased security, time savings, low fuel consumption, and better infrastructure. This system's prominent gadgets

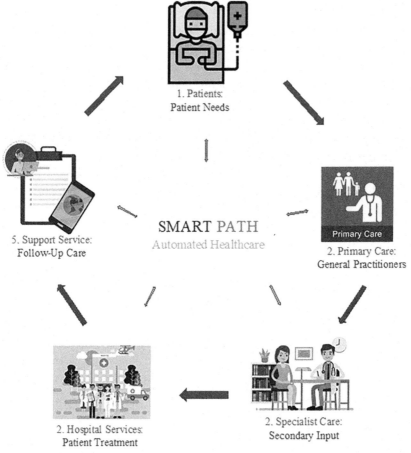

1. Patients:
Patient Needs

SMART PATH
Automated Healthcare

5. Support Service:
Follow-Up Care

2. Primary Care:
General Practitioners

2. Hospital Services:
Patient Treatment

2. Specialist Care:
Secondary Input

Fig. 1 Smart health care.

include infringement tracking and video surveillance, a meteorology condition information service, a motorist alert system, and an automobile com-
munications network, as well as the convenience of rapid and efficient
law enforcement by the authorities and an increase in social welfare. Smart
transportation systems may be classified into four classes, pursuant to the
ERSI understanding of the term:
- To increase transport infrastructure, use dynamic traffic safety
- Service that are dependent on locations
- To enhance traffic dynamics, collaborative traffic efficacy is needed
- Broadband services that are available worldwide
Automobile ad hoc networking is a type of smart transportation system that
is made up of a large number of elevated automobiles [39].

3.4 Smart government

By using ICT convergence for organization, administration, and activities in a thin layer or even across levels, smart government creates benefit for long societal productivity. To put it another way, smart government is the deployment of company operations focused on information and communication technology that triggers informational consistency among governance and the delivery of elevated operations. The ultimate phase in e-government is smart governance [40]. Smart government employs real-time data to prevent violence by boosting spatial awareness, delivering effective and fast disaster reaction, probing crises, and enhancing services to citizens. The key aims of a smart government are depicted in Fig. 2 [41].

3.5 Smart building

Monitoring and grid techniques are used in smart buildings to interact with building systems, report documented electricity use to the smart grid via a smart meter and allow data to be communicated from the smart grid to the facility. These structures will be able to modify their resource patterns periodically dependent on smart energy abilities, as well as enabling property owners to track user their technology [35]. Clearly, the engagement of the smart building with the microgrid results in the achievement of several of the smart grid's primary objectives. Load forecasting, effective communication, optimum reduction, and power transmission are just a few of the many advantages of this relationship. Facility strategic relevance and outside information technology are the two aspects of the smart building ecosystem in particular [42]. Smart economics, safety, transportation, educational, and smart ecosystem elements can all be evaluated in conjunction to the architectural and smart city aspects. [43–45]. Fig. 3 gives the overview of smart building security.

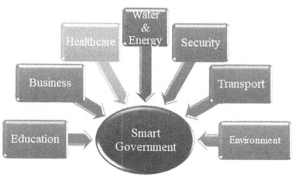

Fig. 2 Smart government.

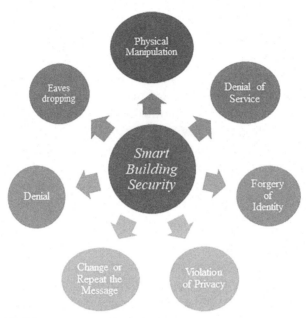

Fig. 3 Smart building.

4. Cybersecurity in smart city planning

4.1 Cybersecurity

When negotiating deals, organizations send confidential material via networks. Nevertheless, cybersecurity provides a range of hardware and software solutions to secure data and communications processing or retention facilities [46]. As the size and diversity of graphical format, so does the need of securing it. Social welfare and financial documents are two of the most important. Cybersecurity refers to the safeguarding of networks, networking, and programs from cyberattacks. Cyberattacks aim to gain exposure to, change, and destroy sensitive data, as well as defraud users and interrupt company activities [47]. Because the number of products has expanded and intruders have gotten more inventive, executing cybersecurity properly and accurately is one of the issues of today's society [48]. Cybersecurity secures computers, networks, cell phones, and communications devices against hostile assaults, and its role may be stated as follows: securing the equipment that individuals use, the contents on these systems, and the identities of the individuals who use this technology [49]. Lacking cybersecurity, businesses are vulnerable to data intrusions and hacker attacks, and they become easy targets for cybercriminals. Eventually, cybersecurity intimidations are bifurcated into three various categories [50]:

(a) Cybercrime: A person or organization that monetizes or sabotages networks.

(b) Cyberattack: Frequently ideological in nature, with the intention of acquiring information for a certain reason.

(c) Cyberterrorism: Intimidates others by damaging computer networks. Malware, SQL injection, phishing, man-in-the-middle attack, denial-of-service assault, and ethical hacking are among the most frequent cyberattacks [51].

4.2 Cybersecurity in smart cities using DL, ML, and AI

To become smart, cities must incorporate new smart technology. Cyberattackers gain fresh opportunities with each technological advances or urban system. Many connections between smart traffic management systems and traffic signals, for example, occur without encrypted or verification, allowing an attacker to manipulate or fake data [52]. Denial-of-service assaults on the smart grid include channel obstruction, cognitive saturation of low-power devices, disseminated denial of access to SCADA, and prolonging a time-critical transmission that might result in widespread outage [53,54]. Falsification of material from numerous sensors is another possible urban hazard; for example, counterfeit sensors to detect seismic events, storms, killings, and other events might result in inaccurate alerts and public alarm [53]. An assailant monitoring on consumer data sent from a smart system to a smart meter poses a serious threat to user security. Furthermore, an adversary faking a client's identity in order to electronically manipulate construction infrastructure might pose a variety of problems for the consumer [55].

Several Internet of things monitors are installed at various data collecting points in a smart city to gather data on congestion, citizen migration, and sewerage, so that the perspectives derived from this data may be used to enhance the operation of productive assets [56]. On data gathered by researchers, DL is commonly employed. The below are some of the most prominent deep learning solutions in smart cities [57,58]:

1. *Metropolitan photorealistic rendering*: Machine learning has been utilized in a number of studies to detect low-income neighborhoods and traffic rates in various parts of cities. Another use of metropolitan fashion is DL-assisted smart space, which finds the best parking spot.

2. *Architecture*: As the foundation of societies, architecture plays a critical role in addressing urban difficulties, and the best decisions may be taken utilizing DL on statistics such as transportation rates, energy usage, and so on. The use of machine learning for Internet scheduling to avoid traffic issues is one of the implementations.

3. *Mobility*: Leveraging the software platform and AI, the mass transit system links people, cars, substructure, and logistics operators. One of its uses is to improve traffic safety in the case of a catastrophe.

4. *Urban administration*: Public organization plays a critical role in evaluating governance regulations and municipal features, as well as assisting in the understanding of the smart city's evolving demands.

5. *Endurance and protracted viability*: Data creation is on the rise. There is a pressing need to build an efficient method for segmenting data that may be used to enhance smart cities. The most significant difficulty is the scarcity of natural commodities. This can be remedied by simulating a smart grid that reduces pollutants and promotes to a higher quality of life. For example, utilizing neural networks, studies have developed a solution for optimum decision-making in the domain of commercial management of waste.

6. *Curriculum*: Big data analysis allows researchers to examine the personality of students in learning. Using machine learning techniques, for example, a novel emotion sensitive method for measuring students' engagement in classes based on their frontal and body language has been presented. Such solutions can address the main problem with digital training, which is the lack of interaction with the instructor.

7. *Health*: Artificial intelligence has cleared the path for the creation of smart health technology using enhanced like DL. Transferring ML algorithms and DL have shown to be quite beneficial in categorization. These strategies, for example, have proven to be more reliable than conventional methods in detecting and forecasting breast cancer.

8. *Protection and anonymity*: Smart cities represent the pinnacle of ICT development. Residents of smart cities are connected via iPhone and synthetic and interconnected technologies such as the Internet of things, resulting in an unfathomable level of convenience and livelihood enhancement.

Despite smart meters, construction equipment, and healthcare systems provide such advantages, they also pose obstacles, such as data protection and confidentiality, keeping information entirety, and preventing illegal access.

DL and related technologies were successful in providing successful remedies for security vulnerabilities involving IoT.

Government have been more concerned about cybersecurity in recent years. The most significant aspects of cybersecurity to which special emphasis should be paid are as follows:

- Identify an unauthorized access.
- Retain a cybersecurity professional to safeguard your credentials.
- Get approval for commercial apps that involve user data.
- To preserve confidentiality, isolate data.
- Avoid disclosing information to the public in order to avoid having to respond to legal entities.
- Document any data deficiencies to the appropriate officials as soon as feasible.
- To safeguard user accounts, use lengthy, complicated credentials with two-factor or inter-verification.
- Never follow links in an email from an unknown sender or on a webpage you are acquainted with.
- It is better not to subscribe to business connections or communicate critical information over unsecured Wi-Fi since data can be intercepted by a hacker.
- Using virus protection is a good idea. This security technology identifies risks and removes them. Consistently keep your security software up-to-date.
- Do not download files sent by unknown senders since they might contain spyware.

Incorporating deep learning techniques into smart cities may go in a number of avenues. Evidently, when the outcomes of the sets and comparable elements of the instruction and assessment material are combined, an instructional technique produces precise conclusions. The next research issue is cognitive transmission, which involves changing or transmitting the instructional and assessment distribution from one platform to another. Researchers might also look at incorporating linguistic frameworks into smart city applications to improve efficacy.

5. Conclusion

Smart city cybersecurity is still in its inception, and numerous policies, structures, programs, and technology solutions are currently being developed in this essential subject. Some studies have given valuable information for policymakers and municipal administrators seeking to better develop and deploy smart city policies and implementation plans, according to an evaluation of the peer-reviewed literature on the topic of smart cities. Other analyses have defined the infrastructure of installing and testing IoT in the smart city to give a platform for experimenting and assessing concepts on a wide scale in real-world situations, but cybersecurity and anonymity have received less attention. Users in the smart city discovered that there was a significant study deficit in this region. Unauthorized provision of information and assaults that aggravate interruption of service access are examples of cybersecurity. With the rise of demographic information technology in municipalities, sophisticated management techniques that leverage the newest technologies and interfaces to make urban functions smarter are becoming increasingly important. Smart cities are a new type of informational and communication technology connectivity. The rate of assaults and exposures will rise as systems are connected and integrated. On the other hand, the more data collected about the whereabouts and actions of global citizens, the less secure anonymity becomes. As a result, it's critical to create solutions that focus on long-term cybersecurity and risk mitigation measures. Resolving these issues, according to the findings of this study, requires the efforts of authorities, equipment and technology makers, and firms that provide IT security services. Furthermore, to avoid significant security events, adaptable systems with strong informational security abilities must be designed, as these occurrences can result in devastating monetary, database, identity, and public confidence losses. Because cybersecurity concerns and threats to users' privacy are not equally significant in the smart city, and appropriate agencies and politicians have restricted resources to address them, subsequent study should include a method for measuring and grading them.

References

[1] M.H. Maruf, M. Asif ul Haq, S.K. Dey, A. Al Mansur, A.S.M. Shihavuddin, Adaptation for sustainable implementation of smart grid in developing countries like Bangladesh, Energy Rep. 6 (2020) 2520–2530.
[2] J. Prasad, R. Samikannu, Barriers to implementation of smart grids and virtual power plant in sub-saharan region—focus Botswana, Energy Rep. 4 (2018) 119–128.
[3] M. Al-Saidi, E. Zaidan, Gulf futuristic cities beyond the headlines: understanding the planned cities megatrend, Energy Rep. 6 (2020) 114–121.
[4] G. Aghajani, N. Ghadimi, Multi-objective energy management in a micro-grid, Energy Rep. 4 (2018) 218–225.
[5] A. Razmjoo, P.A. Østergaard, M. Denai, M.M. Nezhad, S. Mirjalili, Effective policies to overcome barriers in the development of smart cities, Energy Res. Soc. Sci. 79 (2021), 102175.

[6] M. Vitunskaite, Y. He, T. Brandstetter, H. Janicke, Smart cities and cyber security: are we there yet? A comparative study on the role of standards, third party risk management and security ownership, Comput. Secur. 83 (2019) 313–331.

[7] Z.A. Baig, P. Szewczyk, C. Valli, P. Rabadia, P. Hannay, M. Chernyshev, M. Johnstone, et al., Future challenges for smart cities: cyber-security and digital forensics, Digit. Investig. 22 (2017) 3–13.

[8] X. Zhou, S. Li, Z. Li, W. Li, Information diffusion across cyber-physical-social systems in smart city: a survey, Neurocomputing 444 (2021) 203–213.

[9] S. Sengan, V. Subramaniyaswamy, S. Krishnan Nair, V. Indragandhi, J. Manikandan, L. Ravi, Enhancing cyber-physical systems with hybrid smart city cyber security architecture for secure public data-smart network, Futur. Gener. Comput. Syst. 112 (2020) 724–737.

[10] S. Sengan, V. Subramaniyaswamy, V. Indragandhi, P. Velayutham, L. Ravi, Detection of false data cyber-attacks for the assessment of security in smart grid using deep learning, Comput. Electr. Eng. 93 (2021) 107211.

[11] Z. Chen, Application of environmental ecological strategy in smart city space architecture planning, Environ. Technol. Innov. (2021) 101684.

[12] C. Lim, G.-H. Cho, J. Kim, Understanding the linkages of smart-city technologies and applications: key lessons from a text mining approach and a call for future research, Technol. Forecast. Soc. Chang. 170 (2021) 120893.

[13] S. El Hilali, A. Azougagh, A netnographic research on citizen's perception of a future smart city, Cities 115 (2021), 103233.

[14] M. Kashef, A. Visvizi, O. Troisi, Smart city as a smart service system: human-computer interaction and smart city surveillance systems, Comput. Hum. Behav. (2021) 106923.

[15] S.B. Atitallah, M. Driss, W. Boulila, H.B. Ghézala, Leveraging deep learning and IoT big data analytics to support the smart cities development: review and future directions, Comput. Sci. Rev. 38 (2020), 100303.

[16] S.K. Singh, Y.-S. Jeong, J.H. Park, A deep learning-based IoT-oriented infrastructure for secure smart city, Sustain. Cities Soc. 60 (2020), 102252.

[17] A. Belhadi, Y. Djenouri, G. Srivastava, D. Djenouri, J.C.-W. Lin, G. Fortino, Deep learning for pedestrian collective behavior analysis in smart cities: a model of group trajectory outlier detection, Inform. Fusion 65 (2021) 13–20.

[18] L. Liu, Y. Zhang, Smart environment design planning for smart city based on deep learning, Sustain. Energy Technol. Assess. 47 (2021) 101425.

[19] S.M. Nagarajan, G.G. Deverajan, P. Chatterjee, W. Alnumay, U. Ghosh, Effective task scheduling algorithm with deep learning for internet of health things (IoHT) in sustainable smart cities, Sustain. Cities Soc. 71 (2021), 102945.

[20] Y. Liu, W. Zhang, S. Pan, Y. Li, Y. Chen, Analyzing the robotic behavior in a smart city with deep enforcement and imitation learning using IoRT, Comput. Commun. 150 (2020) 346–356.

[21] K. Lee, B.N. Silva, K. Han, Algorithmic implementation of deep learning layer assignment in edge computing based smart city environment, Comput. Electr. Eng. 89 (2021) 106909.

[22] T. Zhang, H. Bai, S. Sun, A self-adaptive deep learning algorithm for intelligent natural gas pipeline control, Energy Rep. 7 (2021) 3488–3496.

[23] N. Lebrument, C. Zumbo-Lebrument, C. Rochette, T.J. Roulet, Triggering participation in smart cities: political efficacy, public administration satisfaction and sense of belonging as drivers of citizens' intention, Technol. Forecast. Soc. Chang. 171 (2021) 120938.

[24] M. Thornbush, O. Golubchikov, Smart energy cities: the evolution of the city-energy-sustainability nexus, Environ. Dev. (2021) 100626.

[25] S. Secinaro, V. Brescia, D. Calandra, P. Biancone, Towards a hybrid model for the management of smart city initiatives, Cities 116 (2021) 103278.

[26] D. Chen, P. Wawrzynski, Z. Lv, Cyber security in smart cities: a review of deep learning-based applications and case studies, Sustain. Cities Soc. (2020) 102655.

[27] S. Barr, S. Lampkin, L. Dawkins, D. Williamson, Smart cities and behavioural change: (un) sustainable mobilities in the neo-liberal city, Geoforum 125 (2021) 140–149.

[28] T. Bjørner, The advantages of and barriers to being smart in a smart city: the perceptions of project managers within a smart city cluster project in Greater Copenhagen, Cities 114 (2021) 103187.

[29] R. Behzad, M. Mehrpooya, M. Marefati, Parametric design and performance evaluation of a novel solar assisted thermionic generator and thermoelectric device hybrid system, Renew. Energy 164 (2021) 194–210.

[30] S. Nakano, A. Washizu, Will smart cities enhance the social capital of residents? The importance of smart neighborhood management, Cities 115 (2021) 103244.

[31] S. Qayyum, F. Ullah, F. Al-Turjman, M. Mojtahedi, Managing smart cities through six sigma DMADICV method: a review-based conceptual framework, Sustain. Cities Soc. (2021) 103022.

[32] V. Kourgiozou, A. Commin, M. Dowson, D. Rovas, D. Mumovic, Scalable pathways to net zero carbon in the UK higher education sector: a systematic review of smart energy systems in university campuses, Renew. Sust. Energ. Rev. 147 (2021) 111234.

[33] T. Ahmad, D. Zhang, Using the internet of things in smart energy systems and networks, Sustain. Cities Soc. (2021) 102783.

[34] W. Huang, M. Marefati, Energy, exergy, environmental and economic comparison of various solar thermal systems using water and Thermia oil B base fluids, and CuO and Al2O3 nanofluids, Energy Rep. 6 (2020) 2919–2947.

[35] M. Mehrpooya, N. Ghadimi, M. Marefati, S.A. Ghorbanian, Numerical investigation of a new combined energy system includes parabolic dish solar collector, Stirling engine and thermoelectric device, Int. J. Energy Res. 45 (11) (2021) 16436–16455.

[36] W. Wang, H. Huang, F. Xiao, Q. Li, L. Xue, J. Jiang, Computation-transferable authenticated key agreement protocol for smart healthcare, J. Syst. Archit. (2021) 102215.

[37] A. Singh, K. Chatterjee, Securing smart healthcare system with edge computing, Comput. Secur. (2021) 102353.

[38] H.H. Jeong, Y.C. Shen, J.P. Jeong, T.T. Oh, A comprehensive survey on vehicular networking for safe and efficient driving in smart transportation: a focus on systems, protocols, and applications, Vehicul. Commun. (2021) 100349.

[39] H. Tao, J. Wang, Z.M. Yaseen, M.N. Mohammed, J.M. Zain, Shrewd vehicle framework model with a streamlined informed approach for green transportation in smart cities, Environ. Impact Assess. Rev. 87 (2021) 106542.

[40] A.T. Chatfield, C.G. Reddick, A framework for internet of things-enabled smart government: a case of IoT cybersecurity policies and use cases in US federal government, Govt. Inf. Q. 36 (2) (2019) 346–357.

[41] J.N. Witanto, H. Lim, M. Atiquzzaman, Smart government framework with geo-crowdsourcing and social media analysis, Futur. Gener. Comput. Syst. 89 (2018) 1–9.

[42] R. Eini, L. Linkous, N. Zohrabi, S. Abdelwahed, Smart building management system: performance specifications and design requirements, J. Build. Eng. 39 (2021) 102222.

[43] J. Al Dakheel, C. Del Pero, N. Aste, F. Leonforte, Smart buildings features and key performance indicators: a review, Sustain. Cities Soc. 61 (2020) 102328.

[44] W. Li, P. Yi, D. Zhang, Y. Zhou, Assessment of coordinated development between social economy and ecological environment: case study of resource-based cities in Northeastern China, Sustain. Cities Soc. 59 (2020) 102208.

[45] M.M. Aborokbah, S. Al-Mutairi, A.K. Sangaiah, O.W. Samuel, Adaptive context aware decision computing paradigm for intensive health care delivery in smart cities—a case analysis, Sustain. Cities Soc. 41 (2018) 919–924.

[46] T. Tam, A. Rao, J. Hall, The good, the bad and the missing: a narrative review of cyber-security implications for australian small businesses, Comput. Secur. (2021) 102385.

[47] S. Hasan, M. Ali, S. Kurnia, R. Thurasamy, Evaluating the cyber security readiness of organizations and its influence on performance, J. Inform. Secur. Appl. 58 (2021) 102726.

[48] M.N. Katsantonis, I. Mavridis, D. Gritzalis, Design and evaluation of COFELET-based approaches for cyber security learning and training, Comput. Secur. 105 (2021) 102263.

[49] B. Uchendu, J.R.C. Nurse, M. Bada, S. Furnell, Developing a cyber security culture: current practices and future needs, Comput. Secur. 109 (2021) 102387.

[50] B. Gunes, G. Kayisoglu, P. Bolat, Cyber security risk assessment for seaports: a case study of a container port, Comput. Secur. 103 (2021) 102196.

[51] J. Huang, D.W.C. Ho, F. Li, W. Yang, Y. Tang, Secure remote state estimation against linear man-in-the-middle attacks using watermarking, Automatica 121 (2020) 109182.

[52] M. Tabaa, F. Monteiro, H. Bensag, A. Dandache, Green industrial internet of things from a smart industry perspectives, Energy Rep. 6 (2020) 430–446.

[53] A. Marahatta, Y. Rajbhandari, A. Shrestha, A. Singh, A. Gachhadar, A. Thapa, Priority-based low voltage DC microgrid system for rural electrification, Energy Rep. 7 (2021) 43–51.

[54] H. Ebrahimian, S. Barmayoon, M. Mohammadi, N. Ghadimi, The price prediction for the energy market based on a new method, Econ. Res.-Ekonomska istraživanja 31 (1) (2018) 313–337.

[55] M. Parasol, The impact of China's 2016 cyber security law on foreign technology firms, and on China's big data and Smart City dreams, Comput. Law Secur. Rev. 34 (1) (2018) 67–98.

[56] A.H. Basori, A.L.B. Abdul Hamid, A.B.F. Mansur, N. Yusof, iMars: intelligent municipality augmented reality service for efficient information dissemination based on deep learning algorithm in smart city of Jeddah, Proc. Comput. Sci. 163 (2019) 93–108.

[57] J. Ashraf, M. Keshk, N. Moustafa, M. Abdel-Basset, H. Khurshid, A.D. Bakhshi, R.R. Mostafa, IoTBoT-IDS: a novel statistical learning-enabled botnet detection framework for protecting networks of smart cities, Sustain. Cities Soc. (2021) 103041.

[58] D. Li, L. Deng, M. Lee, H. Wang, IoT data feature extraction and intrusion detection system for smart cities based on deep migration learning, Int. J. Inf. Manag. 49 (2019) 533–545.

Index

Note: Page numbers followed by *f* indicate figures and *t* indicate tables.

Printed in the United States
by Baker & Taylor Publisher Services